Federal Research on the Biological and Health Effects of Ionizing Radiation

Committee on Federal Research on the Biological and Health Effects of Ionizing Radiation

Division of Medical Sciences

Assembly of Life Sciences

National Research Council

NATIONAL ACADEMY PRESS
Washington, D.C. 1981

NOTICE: The project that is the subject of this report was approved by the Governing Board of the National Research Council, whose members are drawn from the Councils of the National Academy of Sciences, the National Academy of Engineering, and the Institute of Medicine. The members of the committee responsible for the report were chosen for their special competences and with regard for appropriate balance.

This report has been reviewed by a group other than the authors according to procedures approved by a Report Review Committee consisting of members of the National Academy of Sciences, the National Academy of Engineering, and the Institute of Medicine.

The National Research Council was established by the National Academy of Sciences in 1916 to associate the broad community of science and technology with the Academy's purposes of furthering knowledge and of advising the federal government. The Council operates in accordance with general policies determined by the Academy under the authority of its congressional charter of 1863, which establishes the Academy as a private, nonprofit, self-governing membership corporation. The Council has become the principal operating agency of both the National Academy of Sciences and the National Academy of Engineering in the conduct of their services to the government, the public, and the scientific and engineering communities. It is administered jointly by both Academies and the Institute of Medicine. The National Academy of Engineering and the Institute of Medicine were established in 1964 and 1970, respectively, under the charter of the National Academy of Sciences.

The work on which this publication is based was performed pursuant to Contract NO1-CO-95466 with the National Cancer Institute of the U.S. Department of Health and Human Services.

Library of Congress Catalog Card Number 81-84123

International Standard Book Number 0-309-03190-7

Available from:

NATIONAL ACADEMY PRESS
2101 Constitution Avenue, N.W.
Washington, D.C. 20418

Printed in the United States of America

COMMITTEE ON FEDERAL RESEARCH ON BIOLOGICAL AND HEALTH EFFECTS OF IONIZING RADIATION

Members

RUSSELL H. MORGAN, The Johns Hopkins University, Baltimore, Maryland, *Chairman*

ELIE ABEL, Stanford University, Stanford, California

HOWARD BUCKNELL, John Addison Cobb Associates, E. Hampton, New York

JOHN J. CROWLEY, University of Wisconsin, Madison, Wisconsin

PATRICIA W. DURBIN, Lawrence Berkeley Laboratory, Berkeley, California

EDWARD R. EPP, Massachusetts General Hospital, Boston, Massachusetts

PATRICK J. FITZGERALD, Memorial Hospital for Cancer and Allied Diseases, New York, New York

MAURICE S. FOX, Massachusetts Institute of Technology, Cambridge, Massachusetts

HANS E. FRAUENFELDER, University of Illinois, Urbana, Illinois

HARRY K. GENANT, University of California, San Francisco, California

GEORGE T. HARRELL, JR., Pennsylvania State University (Emeritus), Timonium, Maryland

GEORGE B. HUTCHISON, Harvard University, Boston, Massachusetts

LEON O. JACOBSON, University of Chicago Hospitals and Clinics, Chicago, Illinois

JOHN S. LAUGHLIN, Memorial Sloan-Kettering Cancer Center, New York, New York

CYRUS LEVINTHAL, Columbia University, New York, New York

CHARLES W. MAYS, JR., The University of Utah, Salt Lake City, Utah

J. FRANK MCCORMICK, The University of Tennessee, Knoxville, Tennessee

ROBERT D. MOSELEY, JR., The University of New Mexico, Albuquerque, New Mexico

ROBERT D. PHEMISTER, Colorado State University, Fort Collins, Colorado

EDWARD B. ROBERTS, Massachusetts Institute of Technology, Cambridge, Massachusetts

LOUIS ROSEN, Los Alamos Scientific Laboratory, Los Alamos, New Mexico

HARVEY M. SAPOLSKY, Massachusetts Institute of Technology, Cambridge, Massachusetts

CHARLES T. SCHMIDT, Lawrence Berkeley Laboratory, Berkeley, California
RICHARD B. SETLOW, Brookhaven National Laboratory, Upton, New York
JOHN F. SHERMAN, Association of American Medical Colleges, Washington, D.C.
ROY E. SHORE, New York University Medical Center, New York, New York
H. ELDON SUTTON, The University of Texas at Austin, Austin, Texas
JOHN P. WITHERSPOON, Oak Ridge National Laboratory, Oak Ridge, Tennessee
SHELDON WOLFF, University of California, San Francisco, California

Staff Officers
Division of Medical Sciences, National Research Council, Washington, D.C.

DANIEL L. WEISS
ELIZABETH B. HARVEY
ELI J. SALMON
DWAIN W. PARRACK

Editor
FRANCES M. PETER

The committee wishes to acknowledge the contributions of the following staff of the National Research Council:

LUCIANA P. FROST
MARY ANN MEYER
JOAN SEMASINGHE
JEANETTE ANN SPOON
RENEE ST. PIERRE
MARC THOMPSON

CONSULTANTS

EARL J. AINSWORTH, Lawrence Berkeley Laboratory, Berkeley, California
EDWARD L. ALPEN, University of California, Berkeley, California
CLIFFORD C. AMUNDSEN, University of Tennessee, Knoxville, Tennessee
LOWELL L. ANDERSON, Memorial Hospital, New York, New York
ROBERT E. ANDERSON, University of New Mexico School of Medicine, Albuquerque, New Mexico

VICTOR E. ARCHER, University of Utah, Salt Lake City, Utah
F. HERB ATTIX, University of Wisconsin, Madison, Wisconsin
STEPHEN BENJAMIN, Colorado State University College of Veterinary Medicine and Biomedical Sciences, Fort Collins, Colorado
GORDON BLAYLOCK, Oak Ridge National Laboratory, Oak Ridge, Tennessee
MURRAY BOLTON, Massachusetts Institute of Technology, Cambridge, Massachusetts
VICTOR P. BOND, Brookhaven National Laboratory, Upton, New York
DOUGLAS P. BOYD, University of California, San Francisco, California
JAMES BRADBURY, Los Alamos Scientific Laboratory, Los Alamos, New Mexico
JOSEPH D. BRAIN, Harvard School of Public Health, Boston, Massachusetts
A. BERTRAND BRILL, Brookhaven National Laboratory, Upton, New York
ANTONE L. BROOKS, Lovelace Inhalation Toxicology Research Institute, Albuquerque, New Mexico
THOMAS F. BUDINGER, University of California, Berkeley, California
GEORGE W. CASARETT, University of Rochester Medical Center, Rochester, New York
NORMAN COHEN, New York University Medical Center, Tuxedo, New York
ALAN D. CONGER, Temple University, Philadelphia, Pennsylvania
JAMES L. COOLEY, University of Georgia Institute of Ecology, Athens, Georgia
MALCOLM COOPER, University of Chicago Hospital, Chicago, Illinois
COLBERT E. CUSHING, JR., Battelle Pacific Northwest Laboratory, Richland, Washington
LEILA DIAMOND, Wistar Institute, Philadelphia, Pennsylvania
KUNIO DOI, University of Chicago Hospital, Chicago, Illinois
WILLIAM DUMOUCHEL, Massachusetts Institute of Technology, Cambridge, Massachusetts
DONALD L. DUNGWORTH, University of California School of Veterinary Medicine, Davis, California
RALPH E. DURAND, Johns Hopkins Oncology Center, Baltimore, Maryland
JACOB I. FABRIKANT, University of California, Berkeley, California
DONALD J. FLUKE, Duke University, Durham, North Carolina
LESLIE FRALEY, Colorado State University, Fort Collins, Colorado
LEONARD M. FREEMAN, Albert Einstein College of Medicine, Bronx, New York

EDWARD L. GILLETTE, Colorado State University Veterinary Teaching Hospital, Fort Collins, Colorado
CLAYTON S. GIST, Oak Ridge Associated University, Oak Ridge, Tennessee
WALTER B. GOAD, Los Alamos Scientific Laboratory, Los Alamos, New Mexico
MICHAEL GOITEIN, Massachusetts General Hospital, Boston, Massachusetts
MICHAEL S. GOLD, Washington Hospital Center, Washington, D.C.
MARVIN GOLDMAN, University of California, Davis, California
DAVID GOODENOUGH, George Washington University Medical Center, Washington, D.C.
ROBERT GORSON, Stein Research Center, Philadelphia, Pennsylvania
ALEXANDER GOTTSCHALK, Yale University School of Medicine, New Haven, Connecticut
SAMUEL GREENHOUSE, George Washington University, Washington, D.C.
EARLE C. GREGG, Case Western Reserve University, Cleveland, Ohio
PETER GROER, Oak Ridge Associated Universities, Oak Ridge, Tennessee
HIRA GURTOO, Roswell Park Memorial Institute, Buffalo, New York
WAYNE R. HANSEN, Los Alamos Scientific Laboratory, Los Alamos, New Mexico
FRANK HARRIS, Oak Ridge National Laboratory, Oak Ridge, Tennessee
RUSSELL L. HEATH, Idaho National Engineering Laboratory, Idaho Falls, North Dakota
WILLIAM HENDEE, University of Colorado, Denver, Colorado
KURT J. HENLE, University of Utah Medical Center, Salt Lake City, Utah
PAUL HOFFER, Yale University School of Medicine, New Haven, Connecticut
OWEN HOFFMAN, Oak Ridge National Laboratory, Oak Ridge, Tennessee
WILLIAM HOFFMAN, Denison University, Granville, Ohio
ANDREW HUVOS, Memorial Hospital for Cancer and Allied Diseases, New York, New York
A. EVERETT JAMES, JR., Vanderbilt University Hospital, Nashville, Tennessee
HAROLD E. JOHNS, Ontario Cancer Institute, Toronto, Canada
JAMES JOHNSON, Colorado State University, Fort Collins, Colorado
PETER JOSEPH, Columbia Presbyterian Medical Center, New York, New York

MARVIN A. KASTENBAUM, Tobacco Institute, Washington, D.C.
ROBERT KATZ, University of Nebraska, Lincoln, Nebraska
LEON KAUFMAN, University of California, South San Francisco, California
CHARLES A. KELSEY, Cancer Research and Treatment Center, Albuquerque, New Mexico
EDWARD A. KNAPP, Los Alamos Scientific Laboratory, Los Alamos, New Mexico
LEONARD KURLAND, Mayo Clinic, Rochester, New York
LAWRENCE H. LANZL, Rush-Presbyterian-St. Luke's Medical Center, Chicago, Illinois
RICHARD LESTER, University of Texas Medical School, Houston, Texas
EVELYN B. LEVIN, Massachusetts Institute of Technology, Cambridge, Massachusetts
RAY D. LLOYD, University of Utah, Salt Lake City, Utah
ROBERT LOEVINGER, National Bureau of Standards, Washington, D.C.
C. C. LUSHBAUGH, Oak Ridge Associated University, Oak Ridge, Tennessee
JOHN L. MAGEE, Lawrence Berkeley Laboratory, Berkeley, California
KENNETH MANTON, Duke University Center for Demographic Studies, Durham, North Carolina
WILLIAM MEISSNER, New England Deaconess Hospital, Boston, Massachusetts
DANIEL MILLER, Memorial Sloan-Kettering Cancer Institute, New York, New York
CHARLES A. MISTRETTA, University of Wisconsin Medical School, Madison, Wisconsin
THOMAS MITCHELL, Johns Hopkins School of Public Health, Baltimore, Maryland
PAUL MORROW, University of Rochester, Rochester, New York
BRUCE MUGGENBURG, Lovelace Inhalation Toxicology Research Institute, Albuquerque, New Mexico
JERRY S. OLSON, Oak Ridge National Laboratory, Oak Ridge, Tennessee
COLIN G. ORTON, Rhode Island Hospital, Providence, Rhode Island
ROBERT B. PAINTER, University of California, San Francisco, California
DEAN R. PARKER, Austin, Texas
ROBERT G. PARKER, UCLA School of Medicine Center for Health Sciences, Los Angeles, California
HARVEY M. PATT, University of California, San Francisco, California
JAY THOMAS PAYNE, Abbott Northwestern Hospital, Minneapolis, Minnesota

DONALD F. PETERSEN, Los Alamos Scientific Laboratory, Los Alamos, New Mexico
MALCOLM PIKE, University of Southern California School of Medicine, Los Angeles, California
ROBERT E. POLLACK, Columbia University, New York, New York
DAVID RABIN, Georgetown University School of Medicine, Washington, D.C.
HARVEY L. RAGSDALE, Emory University, Atlanta, Georgia
DAVID E. REICHLE, Oak Ridge National Laboratory, Oak Ridge, Tennessee
WILLIAM ROESCH, Battelle Pacific Northwest Laboratory, Richland, Washington
GENEVIEVE ROESSLER, University of Florida, Gainesville, Florida
CHARLES A. ROHDE, Johns Hopkins University, Baltimore, Maryland
DANIEL ROTH, Veterans Administration Medical Center, Walla Walla, Washington
ELIZABETH S. RUSSELL, Jackson Laboratory, Bar Harbor, Maine
LEONARD SCHUMAN, University of Minnesota School of Public Health, Minneapolis, Minnesota
ALLYN H. SEYMOUR, University of Washington, Seattle, Washington
REBECCA SHARITZ, Savannah River Ecology Laboratory, Aiken, South Carolina
ANDREW SIVAK, Arthur D. Little, Inc., Cambridge, Massachusetts
CARL M. SHY, University of North Carolina Institute of Environmental Studies, Chapel Hill, North Carolina
ALFRED SMITH, Los Alamos Scientific Laboratory, Los Alamos, New Mexico
JAMES STEBBINGS, University of Minnesota School of Public Health, Minneapolis, Minnesota
SANDRA TANENBAUM, Massachusetts Institute of Technology, Cambridge, Massachusetts
WILLIAM TEMPLETON, Battelle Pacific Northwest Laboratory, Richland, Washington
CARL F. TESSMER, Olin E. Teague Veterans Center, Temple, Texas
ROY C. THOMPSON, JR., Battelle Pacific Northwest Laboratory, Richland, Washington
LEONARD J. TOLMACH, Washington University School of Medicine, St. Louis, Missouri
JANET TRUBATCH, George Washington University, Washington, D.C.
ROBERT L. ULLRICH, Oak Ridge National Laboratory, Oak Ridge, Tennessee

BURTON E. VAUGHAN, Battelle Pacific Northwest Laboratory, Richland, Washington
GEORGE L. VOELZ, Los Alamos Scientific Laboratory, Los Alamos, New Mexico
ROBERT C. VON BORSTEL, University of Alberta, Edmonton, Canada
SALOME G. WAELSCH, Albert Einstein College of Medicine, Bronx, New York
ROBERT G. WAGGENER, University of Texas Health Science Center, San Antonio, Texas
THOMAS A. WATSON, Ontario Cancer Foundation, Victoria Hospital, London, Ontario, Canada
DAVID WEBER, University of Rochester School of Medicine and Dentistry, Rochester, New York
EDWARD WEBSTER, Massachusetts General Hospital, Boston, Massachusetts
SANFORD L. WEINER, Massachusetts Institute of Technology, Cambridge, Massachusetts
MORRIS J. WIZENBERG, University Hospital and Clinic, Oklahoma City, Oklahoma
PETER WOOTTON, University of Washington, Seattle, Washington
MCDONALD E. WRENN, University of Utah, Salt Lake City, Utah
MARVIN ZELEN, Sidney Farber Cancer Institute, Boston, Massachusetts

Preface

Toward the end of 1979, the National Academy of Sciences (NAS) was asked by the Director of the National Institutes of Health (NIH) to review and evaluate the scope and quality of research on the biological and health effects of ionizing radiation supported or conducted by agencies of the federal government. This request was made in response to legislation (PL 95-622, as amplified by supporting statements in the *Congressional Record* on October 14, 1978) requesting the Secretary of the Department of Health, Education, and Welfare (now the Department of Health and Human Services, DHHS) to develop a comprehensive strategy for research in this field. The legislation mandated that the strategy reflect not only the needs of agencies with obligations to develop new knowledge but also the needs of agencies with responsibilities to protect the public health.

In response to the NIH Director's request, the Committee on Federal Research on Biological and Health Effects of Ionizing Radiation (FREIR) was established within the Division of Medical Sciences, Assembly of Life Sciences, National Research Council. The committee's charge included the following:

- a brief review of the state of knowledge on the biological and health effects of ionizing radiation (Chapters 4, 5);
- a review and evaluation of current research programs in this field (Chapter 9);

• an analysis of the relationship of the research supported or conducted by the several federal agencies to their goals and missions (Chapter 10);

• a critical evaluation of a government-wide agenda for future research into the biological effects of ionizing radiation, which is being developed by an Interagency Research Committee (IRC) (an interim draft of the research agenda was reviewed by the committee, and its critique was delivered to the Director, NIH, for use by the IRC); and

• the identification of scientific studies that need special emphasis to improve the responsiveness of federal agencies to the problems of public health and safety created by ionizing radiation (Chapters 4-8).

The FREIR Committee's review and evaluation of current research involved not only an assessment of the relevant research programs themselves but also an evaluation of the management practices used by the federal government to support these programs. The committee examined the scope and quality of the research programs as well as the quality of the control mechanisms built into the programs, e.g., selection and review processes, planning and execution of research, and coordination among scientists and decisionmakers.

Organization of the Study

Approximately 900 research projects relating to the biological and health effects of ionizing radiation were identified by 15 federal agencies supporting research in this field. This information was further corroborated by questionnaires completed by the principal investigators and by committee and consultant reviews of a representative sample of projects.

To facilitate the review process, the committee classified the research according to main objective and divided the studies into the following seven categories:

• radiation sources and dosimetry;

• medical applications of ionizing radiation and radionuclides;

• control of occupational exposure to ionizing radiation;

• study of transport mechanisms and the effects of radiation and radionuclides on ecological and environmental systems;

• epidemiologic studies of the effects of ionizing radiation on humans;

- laboratory studies of the effects of ionizing radiation on animals, plants, lower forms, cells in tissue culture, and biological substrates; and
- measurements of dose-effect relationships, including the development of models for assessing risk.

No investigations relating to the management of radiation research were identified.

The committee was divided into five subcommittees, each with a specific assignment:

- *Subcommittee I—Overview* coordinated the work of the several subcommittees and blended their findings into this report.
- *Subcommittee II—Medical and Environmental Radiation* examined research on radiation instrumentation, dosimetry, control of occupational radiation exposure, the effects of environmental radiation, and the applications of ionizing radiation in medical diagnosis and therapy.
- *Subcommittee III—Epidemiology* reviewed research on the effects of ionizing radiation in humans.
- *Subcommittee IV—Nonhuman Radiation Effects* examined research on animals, plants, lower forms, cells in tissue culture, and biological substrates.
- *Subcommittee V—Management* studied the management processes used by the several federal agencies supporting and conducting radiation research programs.

Committee Procedure

The review of the research was accomplished in the following manner: Approximately 350 studies were identified as residing within the scope of Subcommittee II; 150 studies within Subcommittee III; and 400 studies within Subcommittee IV. The committee considered all research programs and then selected approximately 150 studies for in-depth, on-site reviews. These studies represented various categories of research that were conducted either intramurally or extramurally. Another 250 were selected for reviews based on submitted written and published information, which was assessed by reviewers and discussed by the subcommittees. All reviews were conducted by committee members and consultants from relevant scientific fields (see list on pp. iii-ix) who had been identified by the committee. In each case, the agency supporting the research, the institution in which the research was conducted, and the project's principal investigator were contacted by staff of the Division of Medical Sciences. They were informed in detail about the purposes of the reviews and were asked to supply background material, such as grant proposals, recent progress reports, and publications. The reviewers were asked to prepare reports describing each project, its aims, and procedures; to provide an analysis of the progress, strengths, and weaknesses of the research; and to comment on the significance of research results. The reviewers who conducted the site visits were asked to describe the relationship of each project to related work conducted at other institutions, to note the special resources available to the research team, and to review the mechanisms for reporting research results and for accounting to the supporting agency.

The reports prepared by the reviewers were circulated to the other members of the review teams for additional comment. Meetings were then held by each subcommittee and its consultants to study the reviewers' reports and to discuss their general conclusions. Each of the several subcommittees then prepared an analysis of each of the research fields under its purview with attention to the following items:

- generic objectives of the projects reviewed;
- quality of the research;
- significance of the research;
- adequacy of the research;
- utilization of the research; and
- conclusions and recommendations of the subcommittee.

Other sources of information were also used by the subcommittees. For example, the Subcommittee on Management conducted approximately 60 interviews with present and past directors and managers of federal research programs, congressional staff members, and representatives of concerned environmental and scientific groups and unions. The interviews covered such matters as agency programs, research utilization, relationships among research programs, and other factors influencing program management. A letter published in the November 16, 1979, issue of *Science* magazine requested comments from members of the scientific community concerning goals that a future research agenda should meet. Approximately 30 letters were received and evaluated. Comments were also collected at an open meeting held in Washington, D.C., on September 15, 1980. The meeting was attended by members of the public as well as by representatives of interested environmental, consumer, industrial, and scientific organizations.

The FREIR Committee also conducted two workshops under the auspices of subcommittees II and IV. These workshops were designed to review current scientific knowledge with respect to the field of radiation biology, the uses of ionizing radiation in medicine, and the control of environmental contamination from radionuclides.

During the course of the FREIR Committee's work, there was much correspondence with federal agencies to determine their methods of identifying research needs, establishing research goals and priorities, evaluating and funding research proposals, reviewing the progress of supported and conducted research, and utilizing research results. The agencies and their representatives were most cooperative. They provided the committee with detailed information that is especially useful as a basis for understanding how the federal agencies perceive their goals and discharge their obligations.

The NIH provided the FREIR Committee with working papers containing extensive information, which served as the background for a public meeting held by NIH to discuss the strategies that might be followed in the development of federally sponsored research in radiation biology. The papers included reviews of current knowledge, identified major issues in each field of research, and outlined the kinds of information that should be developed to overcome deficiencies and uncertainties in the body of scientific knowledge on the biological and human health effects of ionizing radiation.

Seven appendixes have been prepared to supplement the information contained in this report. These have been published in a separate volume, which is also available from the National Academy Press. Appendix A describes the methodology used in this study and lists the projects identified by the committee. Appendixes B, C, and D complement discussions in the text of the main report by providing a more detailed and technical description of the committee's findings in the following areas: epidemiologic studies and other studies of the effects of ionizing radiation in humans, major sources of environmental and medical radiation, and external and internal radiation in animals. Appendix E explains the committee's procedure for exploring management issues affecting the conduct of ionizing radiation research by federal agencies and lists the individuals interviewed in the pursuit of this information. The letter published in *Science* requesting suggestions pertaining to future research on the effects of ionizing radiation appears in Appendix F along with copies of the replies that were received. Appendix G describes the committee's procedure for reviewing research management practices of the federal agencies and contains letters from the agencies describing their activities in the area of research of interest to the committee.

The committee wishes to thank publicly the many scientific consultants who contributed so much to the development of this report. It is also grateful to the many public officials and private citizens who responded thoughtfully and thoroughly to the committee's requests for information, opinion, and guidance. It particularly wishes to thank the scientific investigators who gave so generously of their time and thought. The degree of cooperation obtained from all those who were encumbered with significant demands upon their time and effort bespeaks the extraordinary interest that everyone displayed in seeking to develop a comprehensive and useful document.

It wishes to single out for special thanks Dr. Donald S. Fredrickson, Director of the NIH, and Dr. Charles U. Lowe, Acting Associate Director, Medical Applications of Research, NIH, without whose help and cooperation this study could not have been completed.

The committee is grateful to the staff of the National Research Council, in particular to Dr. Daniel L. Weiss for his untiring assistance in coordinating the work of the committee; to Dr. Eli Salmon, Senior Staff Officer of the committee, and his assistants, Ms. Elizabeth Harvey and Dr. Dwain Parrack; and to Ms. Frances M. Peter, Staff Editor of the report.

It is our hope that the information provided in this report will be useful to the Congress, the federal establishment, and the scientific community in the planning and management of federally supported research on the biological effects of ionizing radiation in the years ahead.

RUSSELL H. MORGAN
Chairman
Committee on Federal Research on
Biological and Health Effects of
Ionizing Radiation

Contents

1 Executive Summary 1

2 The Radiation Sciences—An Overview 11

3 Radiation Quantities and Units 20

4 Human Health Effects of Ionizing Radiation 25

5 Radiation Studies in Animals 36

6 Biological Effects of Ionizing Radiation 64

7 Transport Systems and Ecology 92

8 Radiation Risk Abatement 97

9 Scope and Quality of Current Research 102

10 Research Management 138

11 Future Research and Its Management 164

1
Executive Summary

During the past half century, federally supported research has provided a vast body of knowledge on the biological effects of ionizing radiation. Probably more is currently known about the health risks of ionizing radiation than about any other potentially hazardous agent. As a consequence, there exists a body of scientific information that permits federal authorities to formulate a reasonably conservative and effective system of radiation protection standards and to delineate comprehensive regulatory policies.

As in all scientific disciplines, much remains to be learned. It is therefore important that future research be carefully planned and effectively carried out within the limits of available resources.

Current research constitutes but a small increment of a much larger investigative effort that had its beginnings several decades ago. The committee finds that its quality is generally good. With few exceptions, this research appears to be well conceived and carefully pursued by competent scientists. This is due in no small part to the procedures used by federal agencies to determine that the research objectives and experimental designs of work proposed by their contractors and grantees are appropriate and that the work is carefully and diligently performed. These procedures differ from agency to agency. Some agencies, such as the National Institutes of Health (NIH), use a process of external peer review in which research proposals of a given discipline are reviewed by scientists of similar disciplines. In other agencies, such as the Department of Defense

(DOD), the Department of Energy (DOE), and others, research proposals may be evaluated by agency staff members knowledgeable in the proposal's subject matter and by external peer reviewers. Variations and combinations of these procedures are also used.

Each of the systems has its advantages and disadvantages. However, with the increasing number of federal agencies having an interest in radiation research, the committee believes that there is merit in the adoption of a comparable system of evaluation and review, with a more standardized protocol of how to report results so that they can be combined from several studies sponsored by the different agencies. Of the various systems in use, external peer review seems to be the most objective and provides a means of introducing a broad range of expert scientific guidance to the evaluation and review processes.

Recommendation 1: The committee recommends that federally supported research on the biological effects of ionizing radiation be evaluated within systems of external peer review that are roughly comparable to each other.

The following paragraphs contain the committee's conclusions and recommendations for specific fields of research.

Radiation Dosimetry

In radiation dosimetry, there can be no question about the usefulness of continuing research to improve dosimetric instrumentation. Radiation standards and policies are dependent upon the availability of a broad range of appropriate instrumentation for use by public officials and others involved in the application of ionizing radiation. The committee notes that the quality of dosimetric research in recent years has been exceptional.

From its review of current research, the committee concludes that dosimetric capabilities are reasonably adequate for electromagnetic radiation and charged particles but that dosimetry of neutrons and of mixed radiations requires further development. It also notes that *in vivo* measurements to determine doses from nonuniform distributions of radionuclides deposited within the body should also be improved. Finally, the assessment of radionuclide doses to particular organs, and to specific cells within organs, is in continuing need of further study. Studies of radionuclide uptake, deposition, metabolism, and elimination play an important role in health protection and should therefore be encouraged. Priority should be given to radionuclides to which large populations of humans are exposed.

Recommendation 2: The committee recommends that emphasis be placed upon dosimetric research for neutrons and mixed radiations. Added emphasis should also be placed on the development of dosimetric instrumentation to be used in measurements of nonuniform field distributions of radionuclides. Particular attention should be directed toward measurements of doses to organs and tissues and to specific cells within these organs and tissues.

Epidemiologic Research

In the past, much has been learned regarding the health risks of ionizing radiation from epidemiologic studies of such population groups as the survivors of the Japanese bombings, the uranium mine workers, and several groups of patients in whom x-rays have been used diagnostically and therapeutically. Epidemiologic studies by their very nature extend over long periods. Thus, many of them are still in progress. These studies should be continued with periodic peer review until they have reached their logical conclusions. This may require the protraction of the Japanese studies at least until the end of the life spans of nearly all the irradiated persons and may justify study of subsequent generations assuming that observational techniques now available, or that may be devised, promise to yield new worthwhile information on radiation risks.

Future epidemiologic studies should be undertaken only with great care. From time to time, there will undoubtedly be populations in which exposures to ionizing radiation have occurred and which for various reasons may seem attractive for intensive study. Seldom, however, will these populations be sufficiently large, nor will their radiation doses be well documented and of adequate size to yield statistically significant data on dose-effects relationships and radiation risk.

The committee notes that federal agencies supporting epidemiologic research in recent years have tended to establish their priorities in a manner that is more haphazard than orderly. As a consequence, excessive effort has been directed toward epidemiologic studies of populations exposed to low-dose radiation. Because the results of such studies are likely to be unrewarding, the committee urges that the federal agencies involved in epidemiologic research undertake a restructuring of priorities.

Recommendation 3: The committee recommends that currently supported, large-scale epidemiologic studies on the health effects of ionizing radiation be continued with periodic peer review until they have reached their logical conclusions. Meanwhile, federal agencies supporting epidemiologic research in this field should reexamine their priorities and confine future scientific research to areas that are likely to yield statistically reliable data. The committee recognizes that social and political processes may require responses in the form of surveys and epidemiologic studies even when such efforts are predictably unrewarding scientifically. In such cases, a clear distinction should be made between these studies and those that are scientifically justifiable.

Research on Animals, Lower Life Forms, Plants, Cells in Tissue Culture, and Biological Substrates

Because experience with the effects of ionizing radiation in humans, exposed either accidentally or by design, is necessarily limited, research on animals serving as surrogates for humans and research on animals, lower life forms, cells in tissue culture, and biological substrates to gain an understanding of fundamental radiobiological principles currently constitute the principal avenue of investigation to increase knowledge of the risks to health from exposure to ionizing radiation. Indeed, this is a useful approach to determining dose-response relationships in humans for lifetime dosage levels below 50 rem.

The major health risks following exposure to ionizing radiation include the development of cancer several years later and the development of genetic aberrations or mutations in future generations of exposed individuals. Much can be learned regarding these risks from radiobiological research. However, the fact that such observations are not made in humans raises important questions regarding their applicability in determinations of human risk. Such questions may be expected to disappear only when future research leads to an understanding of the basic principles involved—principles that apply to all living species.

Recommendation 4: The committee recommends that future studies in the field of radiation biology place increased emphasis on an understanding of the mechanisms of radiation carcinogenesis. This is particularly important with respect to carcinogenesis following low doses of low linear energy transfer (LET) radiation. This research should involve cellular and molecular experiments combined with selected studies on irradiated animals and appropriate observations of irradiated human populations. The committee encourages the design and conduct of experiments that test current concepts in models of carcinogenesis in general and radiation carcinogenesis in particular.

The risks of genetic effects from exposure to ionizing radiation have been quantified, to the limits of present knowledge, in the "BEIR III Report" (National Academy of Sciences, 1980). Because of the uncertainty surrounding the nature of the various forms of genetic damage, and their biological consequences, the estimates of risk to humans are imprecise. Consequently, studies in radiation genetics addressing these uncertainties must continue, particularly at the molecular and cellular levels. Furthermore, because reproductive processes such as meiosis and gametogenesis play a large role in the transmission of genetic damage, experimental work is still needed on whole organisms. Additionally, studies will also be required on single cells from animals and plants and on single-celled organisms, especially at very low doses.

Recommendation 5: The committee recommends that future research on radiation genetics place increasing emphasis on resolving the uncertainties surrounding the nature of genetic damage and its biological consequences whether or not radiation is used as a probe of the system. Such research should be directed toward observations not only on single cells from animals and plants but also of whole organisms.

In addition to the need for greater understanding of the basic mechanisms involved in the response of biological systems to ionizing radiation, there is also need for further observations on the responses of whole animals. Such research should be directed toward evaluating the effects of radiation on such subpopulations as developing fetuses, newborn animals, and organisms with special properties influencing their sensitivity to radiation. Physiological and metabolic processes that determine dose distributions from both internal and external radiation sources should be included in these studies.

Recommendation 6: The committee recommends the continuance of research on radiation effects on whole animals, especially studies evaluating these effects in appropriate subpopulations and the physiological and metabolic processes that determine dose distributions in both time and space from internal and external radiation sources.

The raw data generated over the last 35 yr on the delayed effects, both internal and external, of ionizing radiation in animals are the product of an enormous public investment in scientific effort, animals, and money. It has often been recommended in the past that an adequately funded central national archive be established to accommodate this material and make it accessible for continued use. Since many senior investigators in the older projects are approaching retirement age or are being diverted to other work, it is important that their data be retained so that they may be available for use by future scientists. At a time when new studies are few and must be planned with special care, the existence and accessibility of such a data bank would be of great value.

Recommendation 7: The committee urges the creation of an adequately funded central national archive to accommodate the vast amount of raw data on the late radiation effects in animals that have accumulated over the last 35 yr in order to make them accessible for future use.

Environmental Exposure and Radionuclide Transport

In recent years, research directed toward an understanding of the environmental transport of radionuclides has been limited in scope. Following the discontinuation of nuclear weapons testing in the atmosphere, interest in this field appears to have diminished. Although environmental research in the past appears to have provided an adequate basis for the formulation of radiation protection standards and regulatory policies with respect to radionuclide contamination of the environment under normal conditions, there is some question regarding the adequacy of current knowledge regarding accidental, large-scale releases of radionuclides. Because evaluation of recovery from damage to ecological systems from radionuclide contamination requires months and years of observation, it is important that long-term commitments be made to research programs in this field. Moreover, multifactorial experimental work is necessary to identify the additive and synergistic effects in which radionuclide releases are accompanied by other stress factors. There is also a need to develop better models to describe the relationships among radiation source factors, radionuclide dispersal, various biotic processes, and effects. Predictions of these relationships from current models present many uncertainties, largely due to a lack of field validations of these models under various environmental conditions.

Recommendation 8: The committee recommends that long-term, broadly focused research programs be undertaken to increase understanding of the complex transport systems used by radionuclides in a contaminated environment. Supportive research on dietary pathways is especially important. Adequate support should also be given to the continuing development and validation of models by which radionuclide levels within the ecological system may be predicted following radionuclide contamination of the environment.

Occupational Exposure

Levels of occupational exposure to radiation are currently well under the limits now considered to be acceptable. In occupational groups surveyed by DOE and the Nuclear Regulatory Commission (NRC), the average dose per worker has remained essentially constant. This indicates general adherence to present regulations with respect to radiation exposure limits and suggests that risks from occupational exposure are quite low. Although this may be true in general, an exception may be found in certain mining operations where radon levels are difficult to control and many uncertainties exist in regard to the dosage received. *More importantly, critical epidemiologic studies must be pursued in such areas as uranium mining to relate dosage and time of exposure to the development of lung cancer and the significance of other carcinogens such as cigarette smoke, asbestos, and possibly other cocarcinogens.* Another area of insufficient research is chelation therapy, which can improve the removal of radionuclides from internally contaminated workers.

In recent years, efforts to improve dose measurement or to reduce dose for workers in the medical, mining, and nuclear industries have been limited.

Recommendation 9: The committee therefore finds that practical methods to improve occupational dosimetry and to reduce radiation exposure require greater emphasis, and urges attention to vulnerable occupational groups, especially in the mining industries.

Reduction of Risks from Radiation Therapy

During the past two decades, there has been relatively good support for research designed to improve radiation therapy to treat cancer patients. In such treatment, only a portion of the radiation administered to patients is effectively concentrated on the cancerous lesion. Healthy tissues are also irradiated, thereby being subject to potential damage. Consequently, it has been a main concern of the committee to review research directed toward improving the ratio of the tumor radiation dose to the total patient dose. In general, this research has been of good quality, and the committee recommends its continued support.

Radiation-induced carcinogenesis in radiation therapy patients should be evaluated and the treatment modified to reduce the incidence of such carcinogenesis. These evaluations should be both retrospective and prospective and should include not only factors in radiation treatment but also the influence of chemotherapy in combined programs with respect to the initial appearance of cancer.

Recommendation 10: In the field of radiation therapy, research should be directed toward the understanding of radiation-induced carcinogenesis and the further improvement of the tumor:patient dose ratio.

Reduction of Risks From Diagnostic Uses of Radiation

The diagnostic use of radiation in medical practice constitutes the largest source of ionizing radiation to which humans are exposed in the United States (excluding therapeutic radiation, which is administered to a very limited number of people). Because such use of radiation has unmistakable medical benefit, it is important that efforts be made to devise technologies in which benefit is enhanced and risk diminished. In the past, support for such research has been relatively small. To ensure that the best and most cost-effective radiological technologies for dose reduction and improved diagnoses are available to the public, the level of such support should be increased. More adequate funding commensurate with the increase in expenditures for radiological equipment and procedures made by the public and private sectors may be expected to improve the quality and productivity of research by enabling successful teams to continue their work and by raising the probability that worthwhile research proposals will be funded.

Recommendation 11: The committee recommends that special attention be directed toward the development of medical technologies to increase the quantity of diagnostic information derived from these technologies while maintaining or reducing radiation dose. Because the use of radiation in medical practice constitutes a large source of ionizing radiation to which humans are exposed, the committee also recommends the establishment of a focus for management, coordination, and funding of research programs in the medical radiological sciences.

Risk Perception

In addition to scientific information, there are social, economic, and political factors that influence the setting of radiation standards and the development of regulatory programs. Among these is the public's attitude toward radiation risk. The public appears to accept the radiation risks associated with exposure to natural background and, with some concern, those due to medical sources. The risks associated with facilities of the nuclear industry, especially those for generating nuclear power, have raised higher levels of concern. After more than two decades of experience with such facilities in many countries and with a number of system failures including that at Three Mile Island, major segments of the public remain skeptical of their safety.

The processes of selection of acceptable and unacceptable risks by the public are complex and have a substantial bearing on energy policy within the legislative and executive branches of the government. Although the committee makes no specific recommendations in this area, it believes that research conducted to gain a better understanding of the bases upon which risk judgments are made can be useful.

Manpower

A wide range of disciplines is required for research on the biological and health effects of radiation. These include radiation biology, radiation physics, epidemiology, biostatistics, management science, genetics, clinical medicine, and pathology. Manpower requirements have not been systematically studied by this committee with respect to radiation research. Moreover, the broad manpower needs of the required disciplines are not known at this time. However, it is noted that limiting manpower situations may exist in a number of disciplines, for example, in the fields of epidemiology, radiation biology, and ecology.

Recommendation 12: The committee recommends that a study of manpower needs with respect to research on the health and biological effects of radiation be undertaken. The study should be designed to provide information that would clarify the nature and effects of the manpower limitations and suggest specific and interdisciplinary training programs in those areas in which manpower needs are not being met.

Research Management

During the past 10 yr, considerable fragmentation has occurred within the administration of federally supported research on the biological effects of ionizing radiation. Many agencies are now involved in these programs, whereas during the 1950's and early 1960's the great majority of such research was supported by the then-existing Atomic Energy Commission.

This fragmentation does not appear to have been detrimental to the quality or conduct of the research. On the contrary, it may have been beneficial by providing multiple focal points for different interests and emphases and greater opportunities for funding a wide variety of research.

Still, there is a need to coordinate research on the effects of radiation. Gaps in the information base must be identified. Problems common to various agency interests require consideration. Joint resources should be allocated through consultative arrangements. Serious interagency review must be given to major undertakings, especially to the initiation of new epidemiologic studies. The FREIR Committee also recognizes the need to coordinate the regulation of radiation sources.

At present there are two separate committees whose function is to coordinate the development of federally sponsored radiation programs. These are the Interagency Radiation Research Committee (IRRC) and the Radiation Policy Committee (RPC). The committee proposes a consolidation of these two committees, thereby combining their functions.

What seems to be needed now, after the major changes that have occurred during the past decade or so, is an opportunity for the several federal agencies that have been given responsibility for the administration of research on the biological effects of ionizing radiation to carry out their functions without organizational disruption and distraction. If this is done, the radiation research programs of the future will probably be as productive and distinguished as those of the past.

Recommendation 13: Because of the overlapping functions and interdependent relationships of the IRRC and RPC, the committee recommends that they be combined.

In the interests of long-range productivity and the attraction and retention of competent scientists in radiation-related research, attention should be given to the concept of stability and continuity of research programs. This might be accomplished by more frequent awards of 5-yr contracts for both new and renewal projects that peer groups have judged to have a high probability of producing important results. Longer funding cycles have the additional advantage of reducing the need for frequent peer reviews.

Recommendation 14: The committee recommends that in the interests of long-range productivity of research, including that on the biological effects of ionizing radiation and their abatement, more emphasis be placed on the awarding of multiyear contracts and grants.

REFERENCE

National Academy of Sciences. 1980. The Effects on Populations of Exposure to Low Levels of Ionizing Radiation. Report of the Committee on the Biological Effects of Ionizing Radiations. National Academy of Sciences, Washington, D.C. 638 pp.

2

The Radiation Sciences—An Overview

Because the readers of this report may have widely different backgrounds, this chapter and Chapter 3, highlighting the basic principles and definitions of the radiation sciences, have been prepared to assist those not conversant with these subjects to gain a better understanding of the basis for the following chapters concerning the health risks associated with the uses of ionizing radiation and the biological research that is needed to improve risk assessment and abatement. Those familiar with these matters should proceed directly to Chapter 4.

THE MATERIAL UNIVERSE

The fundamental constituents of the universe in which we live are matter and energy, each of which can, under appropriate conditions, be transformed into the other. Examples of matter include the things that we see and feel around us such as the land, the sea, the forests, the sun, and the planets. Examples of energy include electromagnetic radiation such as radio waves, infrared radiation, light, ultraviolet radiation, x-rays, and gamma rays, and the motions associated with matter.

The material world is extremely complex. It is composed of a great variety of physical, chemical, and biological forms. Nevertheless, all of these forms are composed of a series of building blocks whose characteristics exhibit a disarming orderliness. To illustrate, consider what happens when a pure substance, such as a crystal of table salt or sugar, is divided and redivided over and over again into smaller and smaller units. Ultimately, there comes a time when the crystals are no longer recognizable as parts of the substance. At this stage, the crystals of salt or sugar no longer exhibit *any* of the physical or chemical properties of salt or sugar.

The smallest particle into which a substance can be divided and still retain its characteristic properties is called a *molecule,* of which there are many, many kinds. Arranged in an almost endless variety of combinations and configurations, they comprise one of the basic species of building blocks that form the material universe.

If the process of division is carried further, the molecule is found to be composed of a series of even smaller particles, called *atoms.* Like molecules, atoms come in a variety of classes, the properties of each depending on a still smaller series of particles of which they are composed. The initial concepts of atomic structure were first postulated in 1913 by Niels Bohr. He pictured the atom as an infinitesimally small solar system, consisting of a heavy central core or *nucleus* having a positive electrical charge, and a number of light, orbiting particles called *electrons,* each carrying a negative electrical charge. Since that time, the Bohr model of the atom has undergone considerable revision as new knowledge has become available; however, the model is a useful concept for this discussion.

With the exception of the common form of the hydrogen nucleus, atomic nuclei are composed of two types of relatively heavy particles—the *proton,* with a mass of 1.673×10^{-24} grams (g), and the *neutron,* with a mass of 1.675×10^{-24} g. These masses are approximately 1,800 times greater than that of the electron. The proton carries a positive electrical charge that is equal but opposite in sign to the charge on an electron. The neutron has no electrical charge.

The amount of electrical charge carried on an electron is extremely small. It requires a flow of 6.2×10^{18} electrons per second to produce an electrical current of 1 ampere (A), which is approximately the amount of current flowing in a 100-watt (W) lamp that is connected to a conventional 110-volt (V) household power source.

In neutral atoms, the number of positive charges carried on the protons of an atomic nucleus is equal to the number of negative charges carried on the atom's orbital electrons. Because opposite electrical charges attract each other, the electrons are maintained in their orbits by a balance between the electrical forces of attraction toward the nucleus and the centrifugal forces associated with the motions of the electrons.

The chemical properties of an atom are governed by the number of protons (and electrical charges) within the atom's nucleus. In nature, there are 92 species of atoms whose number of nuclear protons range systematically upward from hydrogen, which has a single proton, to uranium, which has 92. Each of these species is called an *element*. In recent years, several elements of greater nuclear size and electrical charge have been produced artificially in nuclear reactors and by particle accelerators. Plutonium-239, with a nucleus of 94 protons and 145 neutrons, is an example of such an atom. The number of protons within the atomic nucleus of an element is called the element's *atomic number*. As the atomic number increases, the ratio of neutrons to protons in atomic nuclei tends to become larger.

An element with a given atomic number may exist in several forms, depending upon the number of neutrons included in the atomic nuclei. For example, hydrogen occurs in three forms: one has a single proton as its nucleus; a second (called deuterium) has a proton and a neutron as its nucleus; and a third (called tritium) has a proton and two neutrons. The various forms of a particular element are called *isotopes* of that element. All exhibit identical chemical properties, but have different atomic weights. The various forms of all atomic species are generally referred to as *nuclides*.

Many nuclides are unstable, undergoing spontaneous nuclear disintegration (*radioactive decay*) during which certain energetic particles and gamma rays are emitted. Such nuclides are called *radionuclides*.

Energy and Ionizing Radiation

The forms of energy of greatest interest in this report are the energies associated with certain charged particles (e.g., electrons and protons) and uncharged particles (e.g., neutrons and electromagnetic radiations).

The particles of electromagnetic radiations are called *photons*. The several forms of electromagnetic radiation (e.g., radio waves and light) differ from one another only with respect to the amount of energy associated with each photon. For radio waves, the amount of this energy is extremely small. As one proceeds through the spectrum from infrared radiation to light, ultraviolet radiation, and, ultimately, to x-rays and gamma rays, the amount of energy becomes progressively greater. For instance, the energy associated with x-rays and gamma rays reaches levels so great that when these radiations fall upon and impart their energy to matter, orbital electrons are ejected from the material's atoms.

An atom from which one or more orbital electrons have been ejected exhibits a positive electrical charge and is called a *positive ion*. The ejected electrons may attach themselves to electrically neutral atoms nearby. These atoms then become negatively charged and are called *negative ions*. Such a process is referred to as *ionization*. Hence, electromagnetic radiation having sufficient photonic energy to cause ionization is called *ionizing radiation*.

Energetic subatomic particles also have the capacity to ionize the matter with which they interact. Therefore, they too are ionizing radiation. Among these are the charged particles emitted during radioactive disintegration or decay.

The production of ionizing radiation involves processes in which forces both within and external to atomic nuclei play a key role. To illustrate, x-rays may be created when energetic electrons interact with the forces prevailing in the extranuclear regions of the atoms of a material on which they impinge. This occurs in x-ray tubes. The radiations emitted during radioactive decay are generated when forces within the nuclei of unstable atoms cause the nuclei to undergo rearrangement with the splitting-off of some of their components.

Naturally occurring radionuclides may emit three types of radiation. One consists of energetic, negatively charged electrons, which are called *beta particles* to distinguish them from energetic electrons produced in other processes. The second consists of helium nuclei, which are composed of two protons and two neutrons. These energetic particles are called *alpha particles*. The third is electromagnetic radiation. Such radiation has been given the special name *gamma radiation* to distinguish it from electromagnetic radiation originating outside the nucleus (x-rays). In each disintegration, one charged particle only is emitted. Gamma radiation is also emitted if the energetics of the particular disintegration scheme requires it. Man-made radionuclides may also include a variety of disintegration schemes in which the emitted charged particle is a fourth type of radiation consisting of *positrons*. These particles have the same rest mass as that of electrons but carry a positive electrical charge.

The energies associated with charged and/or uncharged particles of ionizing radiation are customarily measured in units called *electron volts* (eV). An electron volt is the quantity of energy imparted to an electron when it is moved through an electrical potential of 1 V. It is equal to 1.6×10^{-19} watt-seconds (W-s). X-rays used in medical practice range upward in energy from approximately 10,000 eV (10 keV) to many million electron volts (MeV) per photon. The radiations emitted during radioactive decay also have energies extending over a wide range.

Another type of nuclear disintegration requires mention here because it is a source of many man-made radioactive materials. This is the process of *nuclear fission*, which occurs when certain heavy elements (e.g., uranium-235) are bombarded with neutrons. As a neutron enters and reacts with an atomic nucleus of such an element, the atom splits into two fragments. The sum of the atomic numbers of the two fragments equals the atomic number of the parent atom. Because the ratio of neutrons to protons in atomic nuclei increases with atomic number, several neutrons are left over when nuclear fission occurs and many are set free. Some of these may be captured by other fissionable atoms, and the process is continued in a self-sustaining chain reaction until the number of neutrons and/or the availability of fissionable nuclei drop below critical levels.

The fission fragments are usually radioactive, emitting beta particles and, often, gamma rays.

Every fission is accompanied by the release of substantial amounts of energy (approximately 100 times more than that released during radioactive decay and many million times more than that released during such molecular processes as the burning of coal). Controlled nuclear fission has therefore taken on importance as a means of producing electrical power.

Properties of Ionizing Radiation

Absorption

When ionizing radiation impinges upon matter, some or all of the energy associated with the radiation is transferred to the atoms of the impacted matter in a process called *absorption*. When biological cells or tissues are irradiated, this process sets into motion a series of chemical reactions and biological changes of far-reaching importance.

The physical absorption of energy from ionizing radiation results in two reactions: *excitation* (the elevation of energy levels of orbital electrons without their removal from their parent atoms) and *ionization* (the actual ejection of orbital electrons). As previously stated, removal of an electron from a neutral atom creates a positive ion, whereas the capture of such an electron elsewhere produces a negative ion. Hence, a pair of ions is formed in each ionization event.

On the average, the production of each ion pair and a few excitation events require the absorption of approximately 34 eV of energy in common, noncrystalline materials (e.g., air, water, tissues, etc.).

Most electrons ejected from the atoms of an irradiated volume of matter are sufficiently energetic to induce ionization themselves. Indeed, most ion pairs arise secondarily in this manner. Only a few are produced directly by interactions involving the incident photon radiation itself in so-called primary ionizing events. Charged particles ionize and excite directly.

In tissues, chemical changes may occur in molecules containing both excited and ionized atoms. This activity may be accompanied by the formation of highly reactive intermediates that induce changes in nearby molecules. Nearly all of these reactions take place within a small fraction of a second. In contrast, the biological consequences sometimes require many years to manifest themselves.

When particles of ionizing radiation are absorbed into matter, primary ionizing events are distributed along the radiation's trajectory. For some types of radiation, e.g., helium nuclei and protons, spacing of these events is relatively close. For others, such as x-rays and gamma rays, the events are spaced relatively far apart. The linear rate at which radiant energy is transferred or imparted to matter along a particle's pathway or, stated more simply, the *linear energy transfer (LET)* of radiation, is clearly greater for some types of radiation than for others. The LET of radiation has a strong influence on the extent of the damage produced in biological tissues: the higher the LET, the greater the damage.

Alpha Particles

Alpha particles constitute an example of high-LET radiation. A typical alpha particle, whose energy is 5.5 million eV (5.5 MeV), has a range of only 40 micrometers (μm) in soft tissues (approximately four cell diameters). Moreover, approximately 40,000 ion pairs are produced by such an alpha particle as it traverses a typical cell. Hence, the ionization resulting from the absorption of an alpha particle is concentrated in an extremely small volume and, when it occurs within a cell, may cause severe disorganization of the cell's constituents (such as the strands of genetic material).

Neutrons

Fast neutrons, whose energies range from 10,000 eV (10 keV) to 10 MeV, lose their energy through collision with atomic nuclei. The recoiling atoms, stripped of some of their orbital electrons, create dense tracts of ionization somewhat similar to those of alpha particles. In soft tissues containing large numbers of hydrogen atoms, the recoiling nuclei from neutrons are mainly protons, which can cause severe biological damage.

After a neutron has lost nearly all of its kinetic energy in repeated collisions, it is called a thermal neutron. Typically, the thermal neutron is captured by the nucleus of an atom. With this capture, a gamma ray is often emitted.

Beta Particles

Beta particles (the high-speed electrons emitted from disintegrating atomic nuclei) have a range in soft tissues considerably greater than those of alpha particles of similar energy. For example, a typical beta particle, having an energy of 2 MeV, has a range of approximately 1 cm (about 1,000 cell diameters). Such a particle produces approximately 60 ion pairs while traversing a typical cell. Whereas alpha particles cause intense ionization of a few cells, beta particles typically cause relatively sparse ionization within many cells. This distinction is particularly important if the target material of the cell is concentrated in its nucleus. However, as beta particles and all other charged particles slow down, they produce increasing ion densities, reaching a maximum shortly before the ends of their paths. Hence, low-energy electrons and other charged particles cause the formation of greater numbers of ion pairs per unit length of pathway near the end of their range than they do when they have greater energy.

Electromagnetic Radiation

The absorption of electromagnetic radiation (x-rays and gamma rays) may involve one or more of the following physical processes: photoelectric absorption, Compton scattering, and pair production.

During photoelectric absorption, the energy of an x-ray or gamma ray photon is imparted to one of the orbital electrons of an atom. The electron is instantly ejected with an energy equal to the difference between the energy of the photon and that required to set the electron free. This is the dominant absorption process for photons with energies below 50 keV. Photoelectric absorption increases dramatically as the absorber's atomic number increases. This is the principal reason why the absorption of diagnostic x-rays is substantially greater in bone because of its high calcium and phosphorus content than in soft tissues, which are composed primarily of atoms with much lower atomic numbers.

In Compton scattering, x-rays and gamma rays are deflected by free electrons or by orbital electrons of atoms, giving a *part* of their energy to these electrons, which in turn recoil from the points of interaction. The scattered radiation is thereby reduced in energy but not eliminated during such encounters. The photons that have undergone Compton scattering proceed onward to be rescattered or absorbed elsewhere. Compton scattering is roughly proportional to the density and thickness of the absorbing material. It is affected little by the material's atomic number. Hence, energetic gamma rays from cobalt-60 (1.17 and 1.33 MeV) and cesium-137 (0.66 MeV) are particularly useful in sparing bone from excessive damage during the therapeutic irradiation of tumors in nearby soft tissues.

In electron-positron pair production, the energy of an x-ray or gamma ray photon is converted into an electron-positron pair. The sum of the energies of the electron and positron is equal to the energy of the x-ray or gamma ray photon less the energy equivalence of the rest masses of the electron and positron. After slowing down, the positron combines with a free electron that may be available nearby. Upon combination, the electron and positron are annihilated with the production of two 0.51 MeV photons, emitted in directions nearly opposite to each other (back-to-back). Pair production occurs only near atomic nuclei and when the x-ray or gamma ray photon has an energy equal to or greater than 1.02 MeV, which is the energy equivalance of the sum of the masses of the annihilated electron and positron. This mode of absorption increases slowly as photon energy rises above the 1.02-MeV critical energy level.

In each of the three primary modes of absorption of electromagnetic radiation, the energetic electrons accelerated as a consequence of absorption produce relatively sparse ionization and excitation along their pathways in a manner similar to that produced by beta particles. X-rays, gamma rays, and beta particles are therefore classified as low-LET radiations.

RADIONUCLIDES

Radionuclides comprise one of the major sources of ionizing radiation. Each radionuclide has a unique set of physical properties—the types and energies of the radiations emitted and the rate of decay (usually described by the radionuclide's physical half-life). Moreover, each radionuclide is an isotope of a chemical element and exhibits the element's chemical properties, among them the ability to be oxidized or reduced, to form compounds and radical complexes, and to react with water (hydrolysis). Hence, the chemical properties of a radionuclide determine how it will react with other chemicals in the environment and in the tissues of plants, animals, and humans. In the environment, the chemical properties of a radionuclide also determine whether it will be associated with soil constituents or waters and whether it will be available for uptake by plants and/or animals (freshwater, marine, or terrestrial organisms) and eventually ingested by humans. The chemical properties of an ingested radionuclide determine its ability to penetrate transport systems within the body, the efficiency of its absorption from the point of entry (i.e., the respiratory system, the gastrointestinal system, and/or the skin), its reactions with body fluids and cell constituents, its site or sites of localization within the body, and the manner and rate of its elimination from the body.

3
Radiation Quantities and Units

Since the beginning of their existence, humans have been exposed to ionizing radiation from natural sources. Some of the most important of these are cosmic rays and radiation emanating from radioactive materials within the earth, which often find their way into the food chain and into construction materials.

Ionizing radiation as a useful tool of mankind had its beginnings with the discovery of x-rays by Roentgen in 1895. Immediately following the announcement of this discovery, there was a burst of scientific activity never before equalled. In medicine, the potential of x-rays was recognized at once, and before the turn of the century x-ray equipment was in active clinical use in every corner of the industrialized world. The excitement created by Roentgen's finding was reinforced in 1896 by Becquerel's discovery of radioactivity and by the discoveries of polonium and radium in 1898 by the Curies.

Initially, the hazards of ionizing radiation were unknown. Although a few reports of x-ray "burns" began to appear in the medical literature as early as 1896, the first physicians to use x-rays took few precautions to protect themselves from exposure to this new form of radiation. Many of these pioneers fluoroscoped their hands each day to test their apparatus before their first patients were examined. It did not occur to them that such a practice might be unwise. Before long, the hands of these physicians became inflamed and underwent changes that all too often degenerated into cancer of the skin. After these early experiences, it was soon realized that exposure to ionizing radiation could be harmful and that protective measures should be taken whenever such radiation is used.

Although a general knowledge of the biological effects of ionizing radiation developed rapidly during the early part of this century, research to quantify its effects on living organisms did not begin until the latter half of the 1920's. Such research had to await the development of a system of radiation quantities and units, based on rigorous physical principles, with which radiation levels might be accurately recorded.

Radiation Exposure and the Roentgen

Steps to develop such a system were initiated by a small international group of scientists shortly after World War I. This group proposed the adoption of a unit of *radiation quantity* called the *roentgen* (R), based on the ionization produced by radiation in free air. This unit was defined as the quantity of x- or gamma radiation that produces ions carrying 1 electrostatic unit of either positive or negative charge in 1 cm^3 of air at normal temperature and pressure (i.e., 20° C and 1 atm). Soon, international agreement was reached on the specifications of standard ionization chambers and the roentgen was officially adopted as the unit of radiation quantity. To avoid confusion in terminology, radiation quantity was later renamed *radiation exposure*.

With the completion of these initial steps to place radiation measurement on a sound footing, research on the biological effects of ionizing radiation began in earnest. The first major work was that undertaken by Muller (1927, 1928), who studied the genetic effects of ionizing radiation in fruit flies (work for which he ultimately received the Nobel Prize).

Absorbed Dose and the Rad

With the growth of radiological methods in medicine and the emergence of nuclear industry after World War II, the system of radiation quantities and units soon required further development. It had become apparent that the biological effects of ionizing radiation were related to the quantity of energy absorbed within the exposed tissues and organs. Hence, the concept of *absorbed dose,* defined as the mean energy imparted to 1 g of matter, was introduced, and a unit of absorbed dose, named the *rad*, proposed. To make it numerically similar to the roentgen, this unit was defined as the deposition of 100 ergs of radiant energy* in 1 g of matter.

Dose Equivalent and the Rem

The concept of absorbed dose was a major contribution to the system of radiation quantities and units. However, as the sources of ionizing radiation to which the public might be exposed became more complex, it became necessary to introduce an additional quantity and unit. Certain forms of ionizing radiation, specifically energetic heavy particles such as protons, neutrons, and alpha particles, produce biological effects per unit of absorbed dose that are much greater than those produced by x-rays, gamma rays, and beta particles, i.e., the *relative biological effectiveness* (RBE) of heavy particle radiation is substantially greater than that of light particles and electromagnetic radiations. This should be considered when assessing risks and developing radiation protection standards. Therefore, the system of radiation quantities and units was extended to take this fact into account, and the quantity called *dose equivalent* was introduced.

Dose equivalent is the radiation quantity obtained when absorbed dose is multiplied by a "*quality factor*," which is intended to take into account the relative biological effectiveness of the exposing radiation. The unit of dose equivalent is the *rem*. For low-LET radiations, such as x-rays, gamma rays, and beta rays, the quality factor has been assigned a value of unity. Hence, for these radiations, 1 rem is the equivalent of 1 rad. For high-LET radiations, ICRP Publication 26 (International Commission on Radiological Protection, 1977) recommends quality factors of 20 for alpha particles, 10 for fast neutrons, and 10 for protons. Hence, for these radiations, dose equivalents expressed in rems are numerically much greater than doses expressed in rads. For purposes of radiation protection, radiation levels are now generally expressed in units of dose equivalent (rem) because they reflect more accurately the biological consequences to be expected from exposure to ionizing radiation than do units of absorbed dose. However, since different values for quality factors have been used, the numerical value of the quality factor should always be given when using the rem.

*100 ergs are equivalent to the energy consumed by a 100-W lamp in 10^{-7} s.

Radioactivity and the Curie

Quantities pertinent to radionuclides have also been developed. These include *activity* and *physical half-life*. Activity is the time-rate of disintegration of the atomic nuclei of a radionuclide. The unit of activity is the *curie* (Ci), defined as the amount of a radionuclide in which 37 billion disintegrations occur per second.* This disintegration rate is the activity of 1 g of radium.

Physical half-life is the time required for one-half of the atoms of a radionuclide to decay. When a radionuclide is taken into the body, there is also a biological half-life, which depends on its rate of elimination (excreted or exhaled) from the body. The combination of physical and biological half-lives results in an effective half-life that is smaller than either one.

Fractional Units

Often, radiation levels are so low that it is inconvenient to express radiation exposure, absorbed dose, dose equivalent, and radioactivity in terms of the roentgen, the rad, the rem, and the curie. Therefore, the system has been extended to include fractional units such as the millirad (mrad) and the millirem (mrem). The prefix "milli" indicates that the unit is 1/1,000 of the parent unit. For example, 1 mrem is equal to 0.001 rem; 500 mrem is equal to 0.5 rem. Other widely used prefixes include micro (μ), nano (n), and pico (p) to indicate 10^{-6}, 10^{-9}, and 10^{-12} of the parent unit respectively (e.g., 1 μR 10^{-6} R; 1 nrem = 10^{-9} rem; 1 pCi = 10^{-12} Ci).

The use of fractional units has been the cause of some confusion. A dose of 500 mrem appears to be much larger than a dose of 0.5 rem; and yet, the two doses are exactly the same. Care must therefore be exercised when radiation values, expressed in fractional units, are interpreted.

Gray, Sievert, and Becquerel

Subsequent to an international agreement reached several years ago to apply the metric system worldwide, the International Committee for Weights and Measures adopted a unified set of rules for units

*A curie of a radionuclide emitting beta particles with energies of 1 MeV per particle generates approximately 6 mW of power.

and measurement called the International System of Units (SI). This system is not altogether compatible with the special units developed in the radiation sciences. For example, the SI units of absorbed dose and of dose equivalent are both the joule per kilogram. Because of the certain confusion that would result from the adoption of such units, the International Commission on Radiation Units and Measurement (ICRU) (1980) has requested and obtained a variance with respect to these units. The joule per kilogram when applied to absorbed dose will be known as the *gray* (Gy), and the joule per kilogram when applied to dose equivalent will be known as the *sievert* (Sv). One Gy is equivalent to 100 rad, and 1 Sv is equivalent to 100 rem. The SI unit of radioactivity is the reciprocal second, i.e., one disintegration per second. This unit has been given the special name Becquerel (Bq); 1 Bq = 27 pCi. The SI unit of radiation exposure is the coulomb (C)* per kilogram of air (C/kg). It is approximately 4 $\times$ 10^3 R. The use of the roentgen will be discontinued after a short transition period. No special name has been given the C/kg.

References

International Commission on Radiation Units and Measurements. 1980. Radiation Quantities and Units. ICRU Report 33. International Commission on Radiation Units and Measurements, Washington, D.C. 25 pp.

International Commission on Radiological Protection. 1977. Recommendations of the International Commission on Radiological Protection. ICRP Publication 26. Pergamon Press, Oxford. 53 pp.

Muller, H. J. 1927. Artificial transmutation of the gene. Science 66:84-87.

Muller, H. J. 1928. The production of mutations by x-rays. Proc. Natl. Acad. Sci. USA 14:714-726.

*The coulomb is a unit of electrical charge, approximately equal to the charge carried by 6.25 $\times$ 10^{18} electrons.

4
Human Health Effects of Ionizing Radiation

The effects of ionizing radiation on human health can be classified into two major groups: acute somatic effects and delayed or late somatic and genetic effects. This division is purely pragmatic, depending upon the time an effect is observed, not upon the time at which the effect is induced. Thus, acute somatic effects are restricted to various forms of radiation illness, primarily due to damage of rapidly renewing tissues of the body such as bone marrow and gastrointestinal tract, and they appear within days or weeks after exposures, usually in doses exceeding several hundred rads. The delayed effects, which are the subject of this chapter, can be classified into four major types:

- Cancer may be induced (carcinogenesis) in different tissues and appear after various lengths of time (latent periods) following radiation exposure. The minimal latent periods may vary from 2 yr for leukemia to 15 yr or longer for some solid cancers.
- Genetic or heritable changes (mutagenesis) may occur in the progeny and in future generations derived from exposed humans.
- Teratologic or developmental changes may occur during the development of the embryo or fetus exposed to radiation during gestation.
- Degenerative changes may occur as expressions of local radiation injury, e.g., cataractogenesis, impairment of fertility, and altered immune responses.

Studies in Humans

Ionizing radiation produces its effects by altering the physicochemical structure and molecular dynamics of living cells. When the radiation dose is large and exposure occurs within a short time span, cell death may occur. Smaller doses may affect various cellular physiologic systems, resulting, in the aggregate, in altered cell growth and division, repair capabilities, and other cellular metabolic mechanisms. The critical cellular target of ionizing radiation is the genetic material, i.e., the strands of deoxyribonucleic acid (DNA) organized into chromosomes in cell nuclei. Damage to DNA and chromosomes is considered to be responsible for the most important delayed biological effects of ionizing radiation.

The expression of the effects of ionizing radiation depends upon many physical and biological variables. The physical variables include the dose received, the distribution of exposure over time, e.g., single acute exposure, several smaller acute exposures, or continuous low-level exposure, and the quality of the radiation (high or low LET). The many biological factors include the sex of the irradiated organism, the age at exposure and at the time the disease is expressed, the specific tissue exposed, host susceptibility, and the variability among individuals in their capacity to repair radiation-induced molecular lesions. During the past several decades, much information on the consequences to human health resulting from exposure to ionizing radiation has become available. Most of what is known about human responses to radiation has been obtained from observations made following whole-body or partial-body exposure to acute doses of 50 or more rem. Observations made following smaller acute doses or higher total doses accumulated over long periods are more limited and do not provide adequate reliable evidence of human health effects for interpretation.

The lifetime doses of ionizing radiation to which the general population is exposed are generally less than 25 rem. Humans are continually exposed to natural background radiation from cosmic radiation and terrestrial sources, including that from natural materials in the environment, e.g., materials used for building, and to radioisotopes naturally found within the body. Moreover, in the United States there is an approximately equal exposure to radiation from man-made sources, such as that used in medical and dental diagnostic procedures and pharmaceuticals, color television sets, nuclear power plants, and the combustion of fossil fuels. The combination of natural environmental exposure and exposure resulting from medical activities comprise more than 90% of the ionizing radiation that humans receive in the United States. The average annual dose received by the general public from all sources of radiation is approximately 0.2 rem/yr, or a lifetime average exposure of 15 rem, ranging from 5 to 25 rem for most of the population.

One of the major epidemiologic studies of the effects of ionizing radiation is the study of the delayed health effects appearing in the survivors of the atomic bombings at Hiroshima and Nagasaki. One hundred thousand individuals died either immediately or a few months after the bombings as a result of the blast and heat effects or from acute radiation injury. Since then, among the 284,000 survivors, there have been approximately 430 deaths from malignant tumors in excess of the 69,000 deaths expected from that cause in a comparable population not exposed to the bombings (Beebe *et al.*, 1978). This increment of approximately 0.6% in cancer deaths superimposed on the naturally occurring cancers in the survivors from 1950 to 1974 represents the cancer impact on a population after exposure to acute, whole-body doses of ionizing radiation as high as 600 rem.

It is not practical to obtain direct measurements of the increased risk of cancer induction due to the low level of radiation exposure to which the general public is exposed. To obtain such data with statistical reliability would require the study of huge exposed populations and comparably sized unexposed populations, both observed over periods as long as 30 yr or more (National Academy of Sciences, 1980). It would also require that radiation doses for each individual be accurately recorded. The mitigating effects of dose protraction and confounding influences of other host and environmental effects, such as genetic susceptibility, cigarette smoking, and industrial chemical carcinogens to which the individuals may be exposed, would also have to be known. The magnitude of such a study is great because the number of effects produced by lifetime doses of 25 rem and less is likely to be small compared to the large background of similar abnormalities in the population from causes unrelated to radiation exposure.

Many difficulties are encountered during epidemiologic studies of populations exposed to ionizing radiation. For example, during the early period of the Hanford reactor project, less than 3,000 individuals were occupationally exposed to a dose of 5 rad or greater (mean 15.4 rad) (Gilbert and Marks, 1979). At these doses, approximately 100,000 or more individuals would be required in each of the exposed and unexposed (control) populations to demonstrate a statistically reliable difference in cancer risk, if the risks are no larger than projected in the BEIR III report (National Academy of Sciences, 1980). The sample size will depend upon the selection of dose-response model and end point. During the Three Mile Island nuclear reactor accident, the average incremental exposure to the 2,163,000 people living within 80 km of the reactor was estimated to be less than approximately 1 mrem (Fabrikant, 1981; President's Commission on the Accident at Three Mile Island, 1979). At such a dosage, the population under study would have to be many hundreds of millions, a number that is impractical to study and highly unlikely to have been exposed, for meaningful differences in cancer risks, assuming the BEIR III report dose response projections are reasonable. Moreover, the study would have to continue for 30 yr or more, and the environments of the exposed and unexposed populations would have to remain comparable. It is clear that the requirements of statistical validity impose essentially insurmountable obstacles to the derivation of dose-effect relationships from epidemiologic studies of populations exposed to low doses.

In an attempt to gain insight into the potential health effects of low-level radiation exposure where there are no reliable data on humans, a variety of approaches have been used. Even though direct observations on human populations may be impossible, dose-effect relationships that are demonstrable at high doses in humans can be extrapolated to the low dose range, assuming that one knows how to extrapolate such data correctly, i.e., the radiobiological theory upon which an extrapolation method is posited should conform to observations from experimental studies. For low-LET radiation, such as gamma and x-rays, current basic knowledge of radiobiology is not sufficient to assure the theoretical bases for extrapolation to very low doses. For high-LET radiation, where much less is known, the extrapolation from high to low dose effects is often assumed to be a linear relationship.

The foregoing discussion of extrapolation implies that experimental studies may improve our capability to assess human radiation risk at low doses. However, animal and plant studies require an assurance that we can interpret results for application to humans, an assurance that may be confounded by the innumerable variations among test genera and humans. Cellular, chromosomal, molecular, and submolecular studies should be interpretable in terms of responses in the intact animal, including humans, where a complex of systems for homeostatic control and regulation exist.

As our knowledge of the physicochemical, molecular, cellular, tissue, organismal, and ecological effects of ionizing radiation increases, a concordance among the various types of information is expected to lead to a reasonable foundation for the understanding of effects in humans, at any dose exposure.

Furthermore, such studies will provide a better understanding of the fundamental mechanisms whereby the end results (e.g., tumors, genetic and developmental abnormalities, and degenerative changes) are produced, without respect to any particular causative agent, and should enhance our ability to understand the relationship between the changes and the specific agent, ionizing radiation.

Specific Radiation Effects in Humans

Cancer Induction

The induction of cancer following exposure to radiation has been the subject of the most intensive study of all the effects of radiation. The long-term epidemiologic studies of exposed human populations, such as the Japanese atomic bomb survivors and the patients exposed to therapeutic radiation for treatment of ankylosing spondylitis (National Academy of Sciences, 1980), have been invaluable in estimating risk from radiation exposures. There remains a great deal that should be pursued in studies of these populations to answer questions that cannot be addressed any other way, e.g., the duration of the latent period for specific organs and tissues, the duration of expression of disease and period of increased risk, end results of radiation effects in an entire generation, effects of age and sex, etc.

Genetic Disease

Genetic disease resulting from exposure to ionizing radiation has not been demonstrated in humans. Genetic effects of such exposures have been demonstrated in laboratory animals and are assumed to occur in humans. The marker of a genetic abnormality in animals is a mutation, an alteration of the genetic material that produces an observable effect. Radiation induced mutations are not qualitatively different in their effects from spontaneous mutations or mutations induced by other agents. Mutations are rare, even when they are enhanced by mutagenic agents. The majority of mutations are not expressed in the offspring of irradiated persons. Some may be expressed in later generations as recognizable but infrequent abnormalities.

Experimental methods of detecting mutations in germinal cells before the mutations are transmitted to the next generation would be desirable. For example, a system that could analyze mutations in large numbers of spermatozoa would provide a sensitive indicator of radiation effect. If such a system could be devised, it could equally serve as a risk predictor. Observations in human ova would be equally fruitful.

Moreover, current studies of large populations such as the Japanese atomic bomb survivors should be continued with future generations, assuming that observational techniques under development or those that may be devised, promise to yield worthwhile new information about radiation risks.

Mutagenesis in Somatic Cells

Mutations in somatic cells may be transmitted to cell progeny in the individual bearing the mutation, but not to his or her offspring. There is evidence that cancers may result from somatic mutation, although this view is not universally accepted. Some somatic mutation in cells can be observed directly as changes in the structure of chromosomes. The significance of such alterations for human health is uncertain, although they are generally assumed to indicate an increased risk of detrimental effects.

Measurement of somatic mutation rates is an attractive quantitative approach to defining specific cellular effects of radiation by providing a biological monitor. The cell, rather than the whole individual, becomes the unit of observation. The effect of exposure to a variety of mutagenic agents and their interactions might be studied in a small sample of tissue, e.g., blood (Popp *et al.*, 1979; Sutton, 1971). Such an approach could also permit the identification of individuals with unusual degrees of susceptibility to different mutagenic agents.

The continued development of a variety of methods to detect gross chromosomal abnormalities, point mutations, and DNA sequence aberrations is appropriate. Similarly, evidence of mutational events could be sought by examination of the cell products that are controlled by the genetic constitution of the cell, e.g., enzyme variations, protein changes.

Teratogenesis; Developmental Abnormalities

Exposure of fetuses to ionizing radiation during gestation is of concern because of the high susceptibility of the embryo and fetus to teratogenic effects (National Academy of Sciences, 1980). Relatively little has been published on intrapartum study of the fetus and its environment following exposure to radiation, although opportunities for cytological and chemical analysis by amniocentesis might be available. This type of approach, an attempt to identify induced abnormalities in the microcosm of cells and fluids, requires the development of clear associations between biochemical or cellular variations and overt clinical alterations due to radiation.

Degenerative Changes

Degenerative changes in tissues exposed to ionizing radiation, e.g., cataracts, may result from a combination of mechanisms, including molecular alterations, cell killing, cell injury with secondary inflammatory or immunological responses, or cell alteration by mutation. The process leading to the expression of tissue injury may be long and may include a prolonged latent period during which no overt injurious consequence of exposure is evident. Early identification of the process of change would be useful, as would an understanding of the complex circumstances that determine the quality of the outcome of any particular radiation exposure.

Special Problems and Uncertainties

Dose-Response Relationships

Most long-term effects of radiation in humans have been observed in whole-body exposures in the doses ranging from approximately 100 to 600 rem. Although there is no theory or empirical finding suggesting that there is a dose below which no effect occurs, the problem of estimating the magnitude of such risks in the low (less than 50 rem) dose region is great. Cumulative doses of 30 rem or greater are probably amenable to investigation in human populations by quantitative epidemiologic methods, perhaps supplemented by refinements in the detection of effects. The 30 rem lower limit is only a rough estimate, since the dose at which an epidemiologic study becomes of value depends upon the sensitivity of the health effect being used as a marker, as well as upon the size of the populations under study.

To determine the the relationship of dose to effect at lower doses, a variety of alternative approaches must be used. These are discussed in subsequent chapters of this report.

Dose Rate or Dose Fractionation

Natural background radiation and the majority of occupational and medical radiation exposures are protracted over time, either delivered in fractions or continuously, usually in low doses. Results from radiobiological experiments and very limited observations in humans (National Council on Radiation Protection and Measurements, 1980) suggest that the risk per rad is similar at high doses of highly fractionated radiation, low dose-rate exposure to low-LET radiation, and very low-dose exposure (Boice and Stone, 1978). There are no adequate data in humans from which to address this issue systematically. Priority should be given to studies that have the capability of providing information on dose-rate effects in humans.

Relative Biological Effectiveness (RBE)

The biological effects of different types of ionizing radiation, e.g., gamma radiation, alpha particles, neutrons, etc., depend on the linear energy transfer and on the radiation dose. Therefore, the RBE of the different forms of ionizing radiation affects the nature of the dose-response relationship.

The RBE for the different kinds of radiation at very low doses has been difficult to establish with confidence. There have been few opportunities to make useful observations or reliable estimates of the RBE of high-LET radiations in humans, although human risk coefficients for cancer induction resulting from exposure to high-LET radiations have been evaluated for some sites (Mays, 1978, 1979; Mole, 1980; National Academy of Sciences, 1980, pp. 364-372, 411-420; Rowland *et al.*, 1978; Saccomanno *et al.*, 1976). Advantage should be taken of any appropriate opportunities to investigate RBE of different radiations, especially neutrons and alpha particles, in exposed human populations.

Factors That Modify Radiation Effects

When studying the effects of various levels and types of exposure to ionizing radiation, one must consider the impact of other agents to which the individuals may also be exposed. Relatively little is known about the combined effects of multiple environmental mutagens, including radiation, in humans, although several studies have begun to consider this problem (Archer *et al.*, 1973; Axelson and Sundell, 1978; Boice and Stone, 1978; Ishimaru *et al.*, 1975; Shore *et al.*, 1980b). Efforts to examine the combined effects (additive, synergistic, or ameliorative) of exposure to multiple agents should be considered for existing or contemplated epidemiologic studies, whenever feasible.

Furthermore, relatively little is known about the nature and magnitude of host susceptibility factors on radiation response. Age at time of irradiation affects susceptibility to carcinogenesis (Land *et al.*, 1980; National Academy of Sciences, 1980; Smith and Doll, 1976; Spiess and Mays, 1970). It has been shown that radiation "of the thymus" in children leads to an increase in thyroid cancer. Sex, ethnicity, genetic variation, and other biological risk factors have also been studied to a limited extent (Boice and Stone, 1978; Shore *et al.*, 1980a), but the mechanisms whereby they influence individual suceptibility to the effects of radiation are poorly understood.

Greater knowledge of variations in the susceptibility of humans to the effects of radiation and of the combined effects of radiation and other agents is necessary for a better understanding of the overall impact of radiation on humans.

Dose Distribution

Whereas the foregoing discussion is applicable, in general, to health effects of radiation on humans and to the specific impact of exposure to radiation on the whole body, special consideration must also be given to effects of radiation limited to specific regions, organs, or tissues of the body. Such exposures occur most commonly in medical diagnostic and therapeutic radiation and in individuals occupationally exposed to radionuclides.

In medicine, radiation exposures are usually limited to small fields in which the radiant energy is distributed only to restricted tissues and organs. Although the doses received by such tissues and organs are not difficult to determine, somatic risk, such as cancer induction, is difficult to estimate. Much more information must be acquired before radiation-response relationships, particularly at low doses, are understood and the health risks of limited exposure fields assessed.

Radionuclide Absorption; Internal Emitters

Radionuclides occur naturally in our environment and are produced for industrial and medical applications. They may be absorbed into the tissues of the body following ingestion or inhalation. Hence, they may internally irradiate adjacent cells, tissues, and organs. The radiation effects depend, in part, on the physical characteristics of the emitted radiation and the biological characteristics, including dose distribution due to selective uptake, transport, distribution, metabolism, excretion, and reutilization within the body. Since the physiological behavior of radionuclides is similar to that of their nonradioactive chemical form, their distribution and concentration, and thus their radiation dose, in body tissues and organs can generally be calculated indirectly. An example is the behavior of iodine-131 in the thyroid gland. For certain radioactive materials such as the transuranic radionuclides, little information can be obtained in humans, especially with regard to specific tissue localization and distribution of the emitted energy. To approach the answers to these questions, dependence has been placed upon experiments in animals. This information is important for the maintenance of occupational health and safety in the mining and nuclear power industries and for the general public and workers in nuclear weapons development and testing. As in all experimental studies in animals, the extent to which the information and conclusions can be extrapolated to humans and the confidence that can be placed in such an extrapolation remain uncertain.

REFERENCES

Archer, V. E., J. K. Wagoner, and F. E. Lundin, Jr. 1973. Uranium mining and cigarette smoking effects on man. J. Occup. Med. 15:204-211.

Axelson, O., and L. Sundell. 1978. Mining, lung cancer and smoking. Scand. J. Work Environ. Health 4:46-52.

Beebe, G. W., H. Kato, and C. E. Land. 1978. Studies of the mortality of A-Bomb survivors. 6. Mortality and radiation dose, 1950-1974. Radiat. Res. 75:138-201.

Boice, J. D., Jr., and B. J. Stone. 1978. Interaction between radiation and other breast cancer risk factors. Pp. 231-249 in Late Biological Effects of Ionizing Radiation. Proceedings of the Symposium held by the International Atomic Energy Agency in Vienna, 13-17 March 1978. International Atomic Energy Agency, Vienna.

Fabrikant, J. I. 1981. Health effects of the nuclear accident at Three Mile Island. Health Phys. 40:151-161.

Gilbert, E. S., and S. Marks. 1979. An analysis of the mortality of workers in a nuclear facility. Radiat. Res. 79:122-148.

Ishimaru, T., R. W. Cihak, C. E. Land, A. Steer, and A. Yamada. 1975. Lung cancer at autopsy in A-bomb survivors and controls, Hiroshima and Nagasaki, 1961-1970. II. Smoking, occupation, and A-bomb exposure. Cancer 36:1723-1728.

Land, C. E., J. D. Boice, Jr., R. E. Shore, J. E. Norman, and M. Tokunaga. 1980. Breast cancer risk from low-dose exposures to ionizing radiation: Results of parallel analysis of three exposed populations of women. J. Natl. Cancer Inst. 65:353-376.

Mays, C. W. 1978. Skeletal effects following radium-224 injections into humans. Health Phys. 35:83-90.

Mays, C. W. 1979. Liver cancer risk. Pp. 147-165 in Biological Implications of Radionuclides Released from Nuclear Industries; Vol. II. IABA-SM-237/42. Vienna, Austria.

Mole, R. H. 1980. The radiobiological significance of the studies with radium^{224} and thorotrast. Health Phys. 35:167-174.

National Academy of Sciences. 1980. The Effects on Populations of Exposure to Low Levels of Ionizing Radiation. Report of the Committee on the Biological Effects of Ionizing Radiations. National Academy Press, Washington, D.C. 638 pp.

National Council on Radiation Protection and Measurements. 1980. Influence of Dose and Its Distribution in Time on Dose-Response Relationships for Low-LET Radiations. NCRP Report No. 64. National Council on Radiation Protection and Measurements, Washington, D.C. 216 pp.

Popp, R. A., G. P. Hirsch, and B. S. Bradshaw. 1979. Amino acid substitution: Its use in detection and analysis of genetic variants. Genetics 92 (No. 1, Pt. 1, Suppl.):S39-47.

President's Commission on the Accident at Three Mile Island. 1979. Staff Reports to the President's Commission on the Accident at Three Mile Island. Reports of the Public Health Service and Safety Task Force, p. 9. U.S. Government Printing Office, Washington, D.C.

Rowland, R. E., A. F. Stehney, and H. F. Lucas, Jr. 1978. Dose-response relationships for female dial workers. Radiat. Res. 76:368-383.

Saccomanno, G., V. E. Archer, R. P. Saunders, O. Auerbach, and M. G. Klein. 1976. Early indices of cancer risk among uranium miners with reference to modifying factors. Ann. N.Y. Acad. Sci. 271:377-383.

Shore, R. E., E. D. Woodard, B. S. Pasternack, and L. H. Hempelmann. 1980a. Radiation and host factors in human thyroid tumors following thymus irradiation. Health Phys. 38:451-465.

Shore, R. E., E. D. Woodard, L. H. Hempelmann, and B. S. Pasternack. 1980b. Synergism between radiation and other risk factors for breast cancer. Prevent. Med. 9:815-822.

Smith, P. G., and R. Doll. 1976. Late effects of x irradiation in patients treated for metropathia haemorrhagica. Brit. J. Radiol. 49:224-232.

Spiess, H., and C. W. Mays. 1970. Bone cancers induced by 224 Ra (ThX) in children and adults. Health Phys. 19:713-729.

Sutton, H. E. 1971. Prospects of monitoring environmental mutagenesis through somatic mutations. Pp. 237-248 in E. B. Hook, D. T. Janerich, and I. H. Porter, eds. Monitoring, Birth Defects and Environment. The Problem of Surveillance. Proceedings of a Symposium sponsored by Birth Defects Institute of New York State Department of Health held at Albany, N.Y., Oct. 19-20, 1970. Academic Press, Inc., New York.

5
Radiation Studies in Animals

Knowledge of the biological effects of ionizing radiation in humans is limited to observations made in populations exposed to relatively high doses (50 rem and above). Since data on humans are unreliable at lower exposures, with a few exceptions such as breast and thyroid cancer, it has been necessary to turn to experimentation in animals for information that may be useful in estimating health risks to humans at doses below 50 rem.

Research in animals avoids a number of the problems associated with epidemiologic studies in humans; exposures can be precisely administered, large numbers of subjects can be studied, life spans are shorter, and many other variables can be well controlled. There remains, of course, the critical problem of determining how to extrapolate data obtained from animals to the prediction of risk in humans.

Experimentation in animals may be conveniently divided into two categories. In the first, animals simply serve as surrogates for humans in studies to determine dose-effects relationships. In the second, the experiments are designed to provide a better understanding of the action of radiant energy on biological systems. In this chapter, the extensive body of scientific information developed from animal research is reviewed and evaluated. Because of the huge volume of literature, this review is unavoidably selective rather than exhaustive.

Objectives of Early Radiobiological Studies in Animals

Most of the early animal studies were concerned with improvement of medical practice and advancement of the biological sciences, whether they involved effects of whole- or partial-body radiation from external sources (National Academy of Sciences, 1956, 1961) or the metabolism (absorption, distribution, elimination) of radionuclides and radioactive compounds during growth, good health, and disease (Siri *et al.*, 1949). Between 1896 and 1942, most of the information on the effects of radiation in mammals had been obtained from observations of patients undergoing radiation therapy. Radiotoxicological studies in animals were limited in number and scope, and few, if any, were supported by direct federal grants.

Manhattan Project

Several months before the actual demonstration of a sustained nuclear fission reaction in December 1942, the Manhattan Project Health Group was formed because it was realized that the ranges of energies and intensities of external radiation sources would be dramatically increased in nuclear reactors and that the variety and amounts of radionuclides produced in reactors by fission and neutron-capture would soar almost beyond imagination (Stone, 1951). Protection of the project's work force was the major responsibility of the Health Group. Most of the radiation studies in laboratory animals were undertaken by this group to check the biological assumptions that had been used to establish the radiation standards in existence at that time (Stone, 1951). The Health Group's directors, drawn largely from medical school radiology departments, were aware of the hazards of radiation, and the cases of exposure to radiation were fresh in their minds. These wartime studies of the metabolism and toxicity of the fission products and actinides and other studies on the effects of single or repeated exposures to gamma rays or neutrons were designed to obtain quickly in animals the same kinds of information that had been accumulated slowly and painfully from humans exposed to x-rays and radium.

Protection Standards

In 1942, standards were based in part on the following well established, but semiquantitative observations of radiation effects in therapy patients and radiation workers: large doses of x-rays delivered at high dose rates killed cells; the same or larger doses were better tolerated by normal tissues if the radiation was delivered in fractions several days apart; and larger doses could be tolerated if only part of the body was irradiated. Similar observations were made somewhat later when phosphorus-32 was given internally to treat leukemia in humans (Siri *et al.*, 1949). For radiologists and technicians, it had been recommended that erythema (an appreciable reddening of skin caused by x-rays) and x-ray dermatitis could be prevented if the dose rate was lowered to approximately 1% per month or 10% per year of the acute erythema dose (Mutscheller, 1928; Sievert, 1925). By inference, the depressive effects on the blood-forming and reproductive organs would also be avoided. The "tolerance dose" initially adopted for project workers was 0.1 rem per day (Cantril, 1951). At that time, only one late effect could be related in a rough quantitative way to internal radiation exposure: bone cancer had been observed in radium workers with body contents of more, but not less, than 1 μCi of radium-226 (Evans, 1943). The first standard adopted for workers was set at a body content of 0.1 μCi for radium-226, and for other radionuclides that deposited in bone at the radiation equivalent of a skeletal content of 0.1 μCi of radium-226 + daughters (National Bureau of Standards, 1941).

Animal Studies

By 1946, the Manhattan Project's biologists had completed a number of important studies in animals. Mice and rats were used most often, but some studies included guinea pigs, rabbits, or dogs. Results of these investigations include the following:

- Acutely lethal doses and subacute effects of single and multiple exposures to x-rays, gamma rays, external beta rays, and fast and slow neutrons were determined in several small animals (Blair, 1954; Henshaw *et al.*, 1947; Zirkle, 1951, 1954, 1956).
- Major acute histopathologic effects were determined for x-rays and gamma rays, some important fission products, and alpha-emitting radionuclides (Bloom, 1948; Fink, 1950).
- Effects of ionizing radiation on the formed elements of the blood, the reproductive organs, life spans of rodents exposed continuously to gamma rays at 0.11 to 8.8 R daily were described (Lorenz *et al.*, 1947, 1954).
- These results led directly to reduction of the permissible radiation

dose rate to 0.3 rem per week, because the investigators showed that the margin of safety for continuous low-dose exposures was not as great as had been thought (National Council on Radiation Protection and Measurements, 1959).

- Effects and tolerance levels were determined in several species for a variety of ingested or inhaled uranium compounds (Voegtlin and Hodge, 1949 and 1953).
- The major short-term metabolic properties of soluble forms of the most important fission products and actinide elements were defined after parenteral injection, ingestion, and inhalation (Hamilton, 1947; Scott *et al.*, 1949).
- Some late carcinogenic effects of external and internal radiation sources were demonstrated in mice. Lymphomatous and ovarian tumors were induced by x-rays and neutrons, and bone tumors by strontium-90 and plutonium-239 (Henshaw *et al.*, 1947; Lisco *et al.*, 1947; Zirkle, 1954).

The wartime studies of the effects of radiation in animals were conducted under less than ideal conditions—in haste, with small numbers of poor-quality animals that succumbed early to infections, with primitive radiation monitoring equipment, and with only marginally satisfactory radiation sources. Many studies were terminated before all the animals died. Causes of death were frequently not specifically sought. Apart from death, the end points revealed by classic histopathology were not quantitative. Important information and new leads emanating from these studies included knowledge of life-shortening effects, induction of specific neoplasia, demonstration of the importance of radiation quality, dose, dose-rate, and dose fractionation and protraction; however, none of these subjects were studied quantitatively.

Mammalian Radiobiology After 1946

Many studies in mammalian radiation biology have been federally supported during the 35 yr since the research activities of the Manhattan Project were transferred to the Atomic Energy Commission (AEC). Some of them were designed to meet the needs of the military or to answer practical problems of radiation protection or radiation therapy. Others investigated fundamental mechanisms of the actions of ionizing radiations in whole animals. Whatever the original purposes of the investigations might have been, their results have been mutually supportive.

Investigations of the effects of radiation in animals have produced a vast web of interrelated information. For the purposes of this discussion, the many aspects of research into mammalian radiation biology have been arbitrarily divided into five broad areas: (1) military studies; (2) research on acute radiation lethality and recovery from acute injury; (3) studies of late radiation effects in survivors of acute lethal doses or of animals given protracted exposures—their longevity, incidence of specific diseases, and reproductive and developmental deficiencies; (4) research on modifying factors that affect both the acute and late responses to radiation, including dose, instantaneous dose rate, dose fractionation and/or protraction, and radiation quality (relative biological effectiveness); and (5) studies of radionuclides deposited in the body.

Many of the mammalian radiobiological studies that were initiated and/or supported by the branches of the Department of Defense (DOD) or by the AEC between 1946 and 1956 completed or extended the wartime work. During that decade, and for some years thereafter, results of many of the animal studies, especially from larger projects or from those involving long-lived animals, appeared only in project progress reports, laboratory documents, and proceedings of topical conferences. With the exception of the early progress reports, abstracts of those publications can be found in *Nuclear Science Abstracts*. In the summaries that follow, the cited references are mainly collected papers, conference proceedings, and reviews; they are intended only to be examples and points of entry into the literature.

Military Studies

From the close of World War II to cessation of above-ground tests of nuclear weapons in 1963, branches of the DOD conducted many radiotoxicological studies in animals at military laboratories and weapons test sites for the purpose of extending laboratory studies to military field situations. The major problems addressed in the military studies were acute and late effects in mammals resulting from nuclear detonations and radioactive fallout fields (the gamma rays emitted by the deposited nuclides), inhalation of the nuclides in a passing cloud, and ingestion or inhalation of radionuclides deposited on the ground near the site of the nuclear blast. Some examples are:

- Extension of studies on acute lethality of gamma rays and energetic neutrons to larger animals such as dogs, primates, burros, and farm animals (Bond, 1969; Brown *et al.*, 1968).

• Field studies of a large number of mice (Operation Greenhouse) to determine acute lethality, reduction of life span, and incidence of specific neoplasms after acute exposure to graded doses of gamma rays or neutrons from a nuclear detonation (Upton *et al.*, 1960).

• Field studies in large animals, such as dogs, burros, and sheep, to determine effects resulting from inhalation exposures to a plutonium-239-dioxide aerosol cloud (Stannard, 1973).

• In a military or emergency situation, individuals might be exposed more than once to large radiation doses. Thus, many studies of the "split-dose" design were sponsored by the military to investigate the ability of animals to recover from subacute radiation injury (Dunning and Hilcken, 1958).

Acute Radiation Lethality

The wartime studies of acute radiation effects in animals (Brues and Sacher, 1950; Prosser *et al.*, 1947) provided the foundation for studies initiated during the subsequent 20 yr. The purposes of the new investigations were to characterize completely, in a variety of test animals, the physiology and pathology of acute radiation injury and to find treatments that would improve the chances of survival after large radiation doses (Davidson, 1957; Patt and Brues, 1954a,b). Major efforts were directed toward evaluation of the irreparable component of injury conferred by large doses ($\leq$ 200 rad) of radiation at high dose rates (Sacher, 1958; Sacher and Grahn, 1964).

In the late 1950's, a major technical advance, labeling of the chromosomes of dividing cells with ^{3}H-thymidine, provided a way to study the cell cycle and the renewal rates of tissues. The ^{3}H-thymidine labeling technique was applied to irradiated animals. In less than 10 yr, the cellular mechanisms underlying the major aspects of acute radiation injury were defined (Bond *et al.*, 1965; Hughes *et al.*, 1958; Taylor, 1960), and many individual, strain, and species differences in radiosensitivity were enumerated (Bond and Sugahara, 1969).

Late Effects of Radiation

Manhattan Project investigators identified three major late effects of radiation: shortened life span—not only among survivors of large single doses, but also among those exposed to much smaller daily doses for a long period (Blair, 1954; Henshaw *et al.*, 1947; Lorenz *et al.*, 1947; Zirkle, 1954); induction of specific neoplasia (Henshaw *et al.*, 1947; Law, 1960; Lorenz *et al.*, 1947; National Academy of Sciences, 1961); and reduced fertility and growth rate (Blair, 1954; Zirkle, 1954).

Shortened Life Span and Accelerated Aging In most of the quantitative studies of life-shortening by whole-body irradiation, mice have been selected as the test animal (Grahn and Sacher, 1968; National Academy of Sciences, 1961; Van Cleave, 1968). Radiation-induced life-shortening was found to be related in a complex way to age at exposure, but, for mice irradiated as young adults, the shape of the dose-response curve could be reasonably well defined (National Council on Radiation Protection and Measurements, 1980; Storer *et al.*, 1979). Studies of life-shortening effects in acutely or repeatedly irradiated larger animals, mainly dogs, were begun in the mid-1950's (Andersen and Rosenblatt, 1969; Casarett and Eddy, 1968), and one study of dogs irradiated continuously with low doses of gamma rays is still in progress (Fritz *et al.*, 1978; Norris and Fritz, 1972, 1974; Norris *et al.*, 1976).

Actuarial description of the crude death rates including deaths from all causes suggested that radiation accelerated normal aging in irradiated animals (Brues and Sacher, 1950; Henshaw, 1965; National Academy of Sciences, 1961; Van Cleave, 1968). In the early 1950's, studies were started to investigate the premature aging of animals (mostly rodents) that survived high radiation doses, and a number of techniques—partial marrow shielding, injected spleen or marrow homogenates, parabiosis with an unirradiated partner, delivery of the dose in fractions—were developed to improve survival. Among survivors of single or fractionated high radiation doses, the age of onset was advanced for several neoplasms and a variety of degenerative and atrophic changes that also occurred in aging controls (National Academy of Sciences, 1961; National Council on Radiation Protection and Measurements, 1980; Upton *et al.*, 1960; Van Cleave, 1968). At doses of 200 rad or less, the incidence of leukemia and some solid tumors was still significantly elevated, but, with the exception of atrophy of the mouse ovary and lens, both the incidence and degree of degenerative changes were similar in treated and control animals (Van Cleave, 1968). Thus, the various end points associated with normal aging were not uniformly affected or accelerated by radiation.

The hypothesis that radiation caused such a nonspecific effect as aging was largely abandoned after experimental demonstrations such as that of Lesher (1966), who used ^{3}H-thymidine labeling to demonstrate that the cellular response of rapidly proliferating tissues to radiation was the same for mice under continuous radiation exposure as for their aging controls. When the Operation Greenhouse data, which are frequently cited for their demonstration of radiation-induced aging (Upton *et al.*, 1960), were recalculated by correcting for competing causes of death, cancers were found to be the only specific diseases induced by radiation at low doses (Walburg, 1975).

Selected studies of radiation-induced shortening of life spans and total incidence of tumors and major diseases have been continued because those effects can be investigated as quantitative functions of dose, dose rate, radiation quality, species, age at exposure, and health status with far fewer animals than are needed to establish the quantitative relationships between induction of specific tumors and both radiation and biological variables.

Tumor Induction in Animals Although cancer has long been recognized as a late consequence of exposure to ionizing radiation (Furth and Lorenz, 1954; National Academy of Sciences, 1961), quantitative relationships between the incidence of cancers and doses of external low-LET radiation have been experimentally elusive. The wartime studies in which rodents were irradiated with x-rays, gamma rays, neutrons, or beta particles yielded several important results. They confirmed the high susceptibility of the mouse ovary to tumor induction by radiation, demonstrated that x-rays or neutrons induced lymphomas and mammary tumors in mice, and showed that external beta-particle irradiation induced skin tumors in rats (Henshaw *et al.*, 1947; Lorenz *et al.*, 1947). In the 15 yr that followed, a large amount of research was undertaken to examine the induction of murine leukemias and ovarian tumors by radiation and to determine the dose-response relationships for those neoplasms for acute, high-dose-rate, low-LET radiation exposures (Law, 1960; National Academy of Sciences, 1961).

The scope of quantitative research on radiation-induced carcinogenesis was broadened markedly following publication of the late effects found in 1951 during Operation Greenhouse (Upton *et al.*, 1960). The sample sizes within many groups of exposed mice were large enough and the pathologic examinations sufficiently careful and detailed to demonstrate unequivocally that the nearly instantaneous radiation doses of 200 rads or more increased the incidence and/or advanced the age at onset of a wide variety of both rare and common murine neoplasms in a dose-dependent fashion. Subsequent studies extended those observations to the low dose range and defined dose-response relationships for the widest possible variety of specific neoplasms (Ullrich and Storer, 1979a,b,c; Upton *et al.*, 1969). In these investigations, barrier-sustained, specific-pathogen-free mice were used to eliminate epizootics and reduce losses from chronic infections.

Induction of tumors of the lung, breast, and reproductive organs was demonstrated in female dogs that were exposed to x-rays as young adults and observed over their lifetime (Andersen and Rosenblatt, 1969). However, shapes of dose-response curves could not be determined because the study included only two doses, 100 and 300 rad. A study of the incidence of myeloproliferative disorders and solid tumors in dogs under prolonged gamma-ray exposure is in progress (Fritz *et al.*, 1978).

During the past 10 to 20 yr, many studies have been conducted in special animal models of radiation carcinogenesis. Radiation induction of mammary tumors in female Sprague-Dawley rats was studied originally to define some of the hormonal interactions in breast tumor development. Later, it was used as a model system for investigating such modifying factors as radiation quality, dose, and dose rate (National Academy of Sciences, 1980; National Council on Radiation Protection and Measurements, 1980; Van Cleave, 1968). Quantitative studies have served to identify the critical cells for induction of skin tumors in rats and to examine dose-response relationships for externally applied beta particles, x-rays, and protons (National Academy of Sciences, 1980; National Council on Radiation Protection and Measurements, 1980).

Depression of Fertility in Males Sterilization of small laboratory animals of both sexes by large single exposures to radiation was amply demonstrated before 1940 (National Academy of Sciences, 1961). The Manhattan Project studies and those immediately following systematically examined in males of a number of species the quantitative relationships between radiation dose, dose rate, gonadal weight, microscopic structure, and reproductive performance (Blair, 1954; National Academy of Sciences, 1961; Oakberg, 1968; Van Cleave, 1968; Zirkle, 1954). The demonstration that sperm in adult mammals is produced continuously from a population of stem cells (as are blood cell lines and the intestinal epithelium) and that the various developmental stages of spermatozoa differ in radiosensitivity made it possible to reconcile much of the conflicting evidence on the effects of total dose, dose fractionation, and dose protraction on spermatogenesis (National Academy of Sciences, 1961; Oakberg, 1955).

Even after large single radiation doses, production of spermatogonia by a few surviving stem cells can eventually reinstitute sperm production, but moderate doses applied continuously appear to lead ultimately to aspermia (Blair, 1954; National Academy of Sciences, 1961). The sperm count of the dog had been shown to be a sensitive quantitative measure of damage from chronic exposures to radiation at low dose rates (1 to 10 R per day for 2 yr). Longer exposures at lower doses were investigated during the early 1950's. In these studies, male dogs were exposed to x-rays 5 days per week for the duration of their lives. The lowest dose tested was the permissible occupational rate of 0.3 rem per week. At or below a dose rate of 0.6 R per week, deleterious effects on sperm numbers or motility were not observed (Casarett and Eddy, 1968).

Adult male goats placed in a continuous gamma-radiation field produced very small numbers of sperm after 3 yr of exposure to 7 R per day, but at least one was still siring offspring. Death supervened before all males became permanently aspermic. After 4 yr of exposure to 2.6 R per day, there were approximately 50% fewer sperm and a 50% reduction in motility in the exposed animals as compared to controls, but the reproductive functions of both males and females appeared to be normal (Hupp, 1976).

Depression of Fertility in Females Histologic examination of ovaries from a few irradiated rats and rabbits and a larger number of mice during the Manhattan Project studies revealed that there was a dose-dependent reduction in the numbers of structurally normal primary oocytes and developing follicles and that depletion of cell numbers progressed with time after irradiation (Bloom, 1948). Sterilizing gamma-radiation doses given to female mice were estimated from histologic preparations of ovaries and breeding tests. These doses were independent of dose rate or dose protraction (Lorenz *et al.*, 1947; Zirkle, 1954).

Subsequent studies of radiation-induced reduction of female fertility in several species produced some apparently conflicting results concerning the effects of radiation variables such as total dose, dose rate, dose fractionation, age at exposure, and experimental end points such as degree of infertility, onset of permanent sterility as measured by breeding tests, or gross ovarian structural changes (National Academy of Sciences, 1961; Oakberg, 1967; Van Cleave, 1968). Some of these apparent discrepancies were explained by the demonstration that the pool of female germ cells present at birth is nonmitotic and not replenishable. Thus, it declines to near zero at the end of normal reproductive life. Additional inconsistencies were resolved by the elucidation of the relative radiosensitivity of the stages of oocyte development from embryonic through active reproductive life (Oakberg, 1968).

As an adjunct to and in support of a large research program on the genetic effects of radiation (Oakberg, 1966, 1968), female mice were irradiated once either *in utero* or between birth and sexual maturity. Differential cell counts in serially sectioned ovaries were used to measure radiosensitivity (cell-killing) at the various developmental stages—from primitive oocytes (small follicles) to mature ova. Forced breeding was used to measure total reproductive capacity (Oakberg, 1966, 1968). The frequency distribution of oocyte stages as functions of animal age, radiation dose, and postirradiation interval established that the small oocyte in the mouse and rat is highly radiosensitive, and the more mature stages are relatively resistant (Oakberg, 1968). The high radiosensitivity of small oocytes (resting, nondividing cells) of rodents made these otherwise useful laboratory animals unsuitable for extrapolation to effects of radiation on reproductive processes in the human female.

A project was begun in 1953 to determine the effect of 100 R and 300 R of whole-body x-rays (given in one exposure or up to four weekly fractions) on the ability of sexually mature female beagle dogs to bear and wean two litters before reaching 2 yr of age (Andersen *et al.*, 1961). In 1961, additional animals were irradiated to examine the ability of female beagles (pups, prepuberal, and adults) surviving 300 R of x-rays in one exposure (mean lethal dose) to reproduce continuously through reproductive life (Andersen, 1965). Some aspects of reproduction in the 300-R groups were slightly below the average for controls, but the 100-R group was within normal limits for all measures, including number and timing of fertile matings, litter size, sex ratio, and total weaning weight of pups.

Studies were initiated in farm animals in 1960 to test further the hypothesis that sublethal whole-body gamma irradiation (or high localized doses to ovaries) would have less effect on reproduction in longer-lived animals than in laboratory rodents. In those investigations, results of breeding tests, supplemented by quantitative histology of the ovaries, supported the hypothesis that total reproduction was not measurably impaired, even though oocyte numbers were depressed at doses of 200 R or less (Erickson, 1967; Erickson *et al.*, 1976).

Reproduction was also investigated in young adult female goats exposed continuously to gamma irradiation. At exposures to 7 R per day or less, which permitted significant survival for at least 4 yr in the radiation field, females continued to reproduce and wean young as long as the males in the same radiation field remained fertile (Hupp, 1976). Continuously irradiated female beagles reproduced normally when the dose rate did not exceed 17 R per day (Norris and Fritz, 1974).

Irradiation of Gonads in Utero During fetal life, the germinal cells of both sexes are mitotically active, and, not unexpectedly, the fetal gonads of both sexes are quite sensitive to radiation. Reductions in the number of structurally normal germ cells present after birth and of reproductive capacity of animals irradiated *in utero* have been investigated in several species with emphasis on larger, longer-lived animals and continuous exposure to radiation. Radiation and experimental variables have included total dose, instantaneous dose rate, dose protraction, and stage of fetal development in studies on goats (Hupp, 1976; Sikov and Mahlum, 1969), pigs (Erickson and Martin, 1976), cows (Erickson and Reynolds, 1978), and monkeys (Andersen *et al.*, 1977).

Irradiation of Embryo and Fetus During the 1930's, descriptive studies were conducted on the susceptibility of embryos, mainly of invertebrates, birds, or amphibia, to radiation injury, and radiation was first used as a tool in mammalian embryology (Russell, 1954). By the mid-1950's, major advances had been made in describing the effects of radiation on the prenatal mammal, especially the mouse and rat, and many fundamental relationships had been established between experimental variables such as radiation dose and gestational stage at time of exposure and such biological end points as fetal death, structural abnormality, and retardation of growth (Russell, 1954). Concise summaries have been written for the dose-related and stage-related congenital malformations induced by radiation in fetal rodents (Van Cleave, 1968).

Because continuous irradiation was considered to be the most likely pattern of human exposure after a nuclear war, there was a perceived need to obtain information on the effects of radiation in animals exposed continuously from conception to birth or to death. The first such investigations of fetal survival, incidence of malformations, postnatal growth, life span, and reproductive capability of continually irradiated populations were conducted in rats (Coppenger and

Brown, 1967). Later, some small-scale studies were conducted in dogs (Norris and Fritz, 1974) and in goats (Hupp *et al.*, 1965).

The mouse, rat, hamster, and rabbit exhibited different sensitivities to radiation resulting in death *in utero* or production of malformed young, depending upon the stages of gestation at the time of irradiation (Phemister *et al.*, 1969). To assess these effects in animals with longer lives, several programs using dogs, pigs, cows, and goats were begun in the 1960's. The rates of fetal passage through the various gestational stages are slower for these animals, and the degree of maturity at birth is greater than for rodents (Sikov and Mahlum, 1969).

The effects observed at or soon after birth in offspring irradiated *in utero* are largely associated with high doses and high dose rates. Most effects that have been studied quantitatively are dose-dependent and can be explained by the loss of critical cells through the mitotic failure. Such cells cannot be replaced efficiently within the time and space imposed by the biological constraints inherent in intrauterine development. With the cessation of above-ground nuclear weapons testing in the United States and the beginnings of development of commercial nuclear power, research attention shifted largely to the long-term effects on large populations of nearly continuous low-dose, low-dose-rate radiation. Epidemiologic studies of perinatally irradiated human populations suggest an increase in the rate at which neoplasia occurs before maturity (National Academy of Sciences, 1980). Experimental verification was sought in a major investigation begun about 1970. In this study, which is still in progress, investigators are irradiating beagle dogs before birth, soon after birth, and during adulthood to compare the sensitivity for cancer induction in a large, fairly long-lived mammal (Thomassen *et al.*, 1978).

Physical Factors in Radiation Dose-Response Relationships

Dose and Dose Rate

Quantitative descriptions of the relationships between total dose for single exposures to low-LET radiation at high dose rates and acute lethality or other rapidly appearing manifestations of cellular damage were early research goals in mammalian radiobiology (Bloom, 1948; Evans, 1952; Patt and Brues, 1954a) To answer the question, "How much radiation is safe?", studies were started in the early 1950's to define the dose dependence of late radiation effects, particularly those shortening life and inducing cancer (Grahn and Sacher, 1968; National Council on Radiation Protection and Measurements, 1980; Upton *et al.*, 1960; Van Cleave, 1968). More recent studies have been reported by Storer *et al.* (1979) and Ullrich and Storer (1979a,b,c). Studies in animals exposed to wide ranges of doses have yielded data that can be applied when examining the quantitative relationships between radiation risk and dose magnitude for low-LET radiation. Such applications have been analyzed in a comprehensive review by the National Council on Radiation Protection and Measurements (1980).

The planning of radiation therapy and the development of strategies for basic protection have long rested on the qualitative observation that acute radiation effects (per unit dose of low-LET radiation) are substantially reduced by dose fractionation, during which exposures to portions of the total dose alternate with radiation-free periods. Quantitative investigations of the modification of mammalian response to radiation dose caused by fractionation and/or protraction were begun during the Manhattan Project (Blair, 1954; Henshaw *et al.*, 1947; Lorenz *et al.*, 1947; Zirkle, 1954). In many studies of radiation dose-response for both acute and late end points, graded total doses were given at different dose rates or fractionation patterns. The radiation variables have been described by Grahn and Sacher (1968). *Dose intensity* (or instantaneous dose rate) is the minute or hourly rate at which radiation is delivered by the source. *Dose rate* is a more general term applied to the average dose absorbed by the animal per unit of time, such as rad per day or rem per year. *Fractionation* is division of the total dose and its delivery at intervals. A small number of closely spaced fractions approximates a single, acute exposure for many late radiation effects. *Interval between fractions* is the radiation-free period between doses. It does not exist for continuous exposure. *Protraction period* is the total time over which the dose is delivered, regardless of the number of fractions or the instantaneous dose rate. This term is usually used to describe an irradiation period that covers a significant portion of the life span.

Investigators have studied the extremes of the rate-related variables, ranging from a single high-intensity exposure to gamma rays from a nuclear detonation during Operation Greenhouse (Upton *et al.*, 1960) to lifetime exposures at very low dose rates approaching natural background, e.g., 0.11 R per 8-hr day (Lorenz *et al.*, 1947, 1954) and 0.06 R per day, 5 days per week (Casarett and Eddy, 1968). Grahn *et al.* (1978) and the National Council on Radiation Protection and Measurements (1980) have analyzed and reviewed the major completed mammalian studies in which dose-rate variables were investigated in mammals over a range of total doses. Several investigations of quantitative dose-response for late effects, in which dose rate and fractionation and/or protraction are major variables, are either in progress or have not yet been completely reported. These include studies in mice (Ainsworth *et al.*, 1976; Spaulding *et al.*, 1978) and in dogs (Fritz *et al.*, 1978).

Radiation Quality

Measurement of the relative biological effectiveness (RBE) of radiations of different quality, i.e., those that produce different numbers of ion pairs per unit of path length in matter, has long been a major research goal in mammalian radiobiology. The purposes for measuring RBE's are both theoretical, to understand the mechanisms of action of different radiations on cells, and practical, so that protection standards can be set for radiations for which there are no direct data from experience in humans. Most RBE's have been measured by using acute biological end points such as lethality and depression of body or organ weights or blood cell counts (Evans, 1952; Schambra *et al.*, 1967; Storer *et al.*, 1957; Tobias, 1950).

The underlying mechanisms for injury, mutation, and repair in irradiated cells do not appear to be entirely comparable for radiations of greatly different LET (neutrons and alpha particles compared to beta particles and gamma rays), and accumulating data indicate not only that RBE's differ for acute and late effects but that they are also functions of dose and dose rate. The relative effectiveness of some internally deposited alpha- and beta-emitting radionuclides for specific late end points, e.g., bone and lung cancer, has been studied when suitable isotopes or comparably metabolized chemical forms have been available (Finkel, 1959; Mays and Finkel, 1980; Sanders *et al.*, 1970). The relative effectiveness of both single and protracted exposures to neutrons and gamma rays has been investigated using life-shortening and specific cancer induction as biological end points (Darden, 1969; Storer and Sanders, 1958; Storer *et al.*, 1958, 1979; Ullrich and Storer, 1979a,b,c; Upton *et al.*, 1960). At least one of these projects is still in progress (Ainsworth *et al.*, 1976).

Radionuclides Deposited in the Body

The listing that follows is not exhaustive. It is intended to present in a roughly chronological sequence important studies of radionuclide toxicity in animals that have been conducted since 1946 under federal sponsorship and the reasons for initiating them.

• By 1956, metabolic studies of radionuclides had been completed (at least in rodents) for most of the elements of the periodic table and the man-made transuranium elements (International Commission on Radiological Protection, 1959).

• Dosage-response studies of osteosarcoma induction in mice by bone-seeking radionuclides, which were started during the Manhattan Project, were completed for radium-226, plutonium-239, strontium-90, calcium-45, and several uranium isotopes (Finkel, 1959). Many similar studies, including multiple injections and continuous feeding in mice, were completed somewhat later for strontium-90. Reticular tissue tumors were also induced by continuous feeding of strontium-90 to mice (Finkel *et al.*, 1960).

• In 1950 dosage-response studies of bone-seeking radionuclides were initiated in beagle dogs, specifically to provide better support for plutonium standards and to verify the results from rodent tests in a larger, longer-lived animal. The initial study was designed to replicate the exposure of humans to radium and thorium isotopes in radium dial paint. Investigators used graded dosages (soluble compounds in a single parenteral injection) of several isotopes of radium, thorium-228, and plutonium-239 (Dougherty *et al.*, 1962). Strontium-90 was added to these tests in 1955 because of concern about long-term effects of worldwide fallout from weapons tests. In 1966, americium-241 was added because of its importance as a byproduct of nuclear reactors. Radium-224 was studied beginning in 1977 to duplicate in an animal the exposures of humans to a short-lived radium isotope, during which much of the alpha energy is absorbed by bone surface cells instead of by minerals in bone (Mays, 1978; Mays *et al.*, 1969). Significant numbers of bone tumors were induced at all injected dosages of radium-226 and plutonium-239 in the original testing. Subsequent tests with intermediate and lower dosages are still in progress (Stover *et al.*, 1972). Wrenn (1979) has listed annual reports of these studies and provided tables of injections used in the test animals.

• All but a few compounds of natural uranium have been shown to be absorbed into the body after inhalation, and their chemical toxicity in the kidneys was found to be limiting (Voegtlin and Hodge, 1949 and 1953). However, uranium dioxide was retained in the lung as a long-lived radiation source. Dogs and monkeys were exposed to uranium-238-dioxide dust daily for 5 yr (Leach *et al.*, 1970). During the subsequent 5 yr, after the exposure ended, lung tumors were observed in some dogs, but not in the monkeys (Leach *et al.*, 1973). A comparable study of inhaled thorium-232-dioxide dust was started, but not continued long enough to produce useful results.

• Iodine-131 was recognized early as a major health hazard of reactor operations and fuel reprocessing. In a large program started in 1950, ewes and growing lambs were fed graded dosages of iodine-131 continuously for several years. Chronic dose-dependent hypothyroidism, rather than thyroid neoplasia, was the major late effect (Bustad, 1963; Bustad *et al.*, 1957).

• For approximately 10 yr immediately after World War II, a large number of studies conducted in three laboratories examined the distribution, retention, and acute and late biological effects of polonium-210, a short-lived alpha-emitter. Distribution was observed largely in soft tissues, and dose-related late effects were, chiefly, degeneration and atrophy of soft tissues and tumors of renal, reticular, and connective tissues (Fink, 1950; Finkel, 1959; Stannard and Casarett, 1964).

• Preliminary studies of beta- and alpha-emitting radionuclides instilled or implanted in the lungs of rodents induced lung tumors (Sanders *et al.*, 1970). Inhalation during chemical reprocessing was the most common form of occupational plutonium exposure. In 1955, investigations were begun to examine the effects of plutonium inhaled by beagle dogs exposed to plutonium dioxide, plutonium tetrafluoride, and plutonium nitrate, the compounds encountered most often occupationally (Bair, 1970). In the original experiment, the dogs receiving the lowest dosage level of plutonium dioxide eventually developed lung tumors. Later, several lower dosage levels were added. Compounds of plutonium-238 have now also been included (Bair *et al.*, 1973; Drucker, 1980). This study is still in progress.

Because of the long physical half-life, long retention in bone, and great geochemical and biological mobility of strontium-90, its presence in worldwide fallout from atmospheric weapons tests was the stimulus for initiating chronic feeding studies in beagle dogs, begun in 1956, and in miniature swine, beginning in 1959. Bone tumors were anticipated as an end point, but myeloproliferative disease (a form of myelogeneous leukemia) was unexpectedly induced in both species when strontium-90 was fed continuously during growth (Clarke *et al.*, 1968; Goldman *et al.*, 1972). A series of multiple injections of radium-226 was included in the dog experiment to link that project both to the studies of single strontium-90 and radium-226 injections in beagle dogs and to the human experience with radium dial paint. This study is not yet complete.

About 1960, studies of the acute and late effects of single injections of two other major fission products, cesium-137 and cerium-144, were begun in beagle dogs. As had been found for high dosages of strontium isotopes and for external radiation, the major cause of death from high injected dosages of these isotopes was bone marrow aplasia. Cerium-144 induced bone tumors as expected, but hepatic degeneration and tumors of nervous tissues were unexpected findings in the survivors of high dosages of cesium-137 (Fritz *et al.*, 1966; National Council on Radiation Protection and Measurements, 1977). These studies have not been extended to continuous intakes or lower dosages.

Another major program was begun in 1960 to investigate the deposition, fate, and biological effects of inhaled fission products. Several years of developmental effort were required to perfect measurable and reproducible exposure and monitoring methods for administering high levels of radioactive materials by inhalation. Then, in large-scale investigation, several prototype radionuclides (in soluble or insoluble forms and of controlled, known particle size distributions) were administered by nose-only inhalation to beagle dogs (International Commission on Radiological Protection, 1959; Sanders *et al.*, 1973). Development of a technique for preparing and separating insoluble aerosol particles by size, followed by resuspension and reaerosolization, made it possible to undertake investigations in rodents and dogs to examine the deposition and late effects of alpha-emitting particles having greatly different sizes, numbers, and specific radioactivity (Raabe *et al.*, 1975). For listings of project reports and exposure tables, see Henderson *et al.* (1979). This project is continuing.

It has been known for many years that lung cancer is a major late effect of underground uranium mining, but early experiments of radon inhalation in animals were inconclusive (Holaday, 1973). Cancer incidence appeared to be related to inhalation of the short-lived daughters of radon, but the epidemiology was confounded by the presence of uranium ore dust and diesel fumes in mine air and by cigarette smoking. In a study begun in about 1960, beagle dogs were exposed by inhalation to mine air constituents and cigarette smoke singly or in combinations (Cross, 1978). Some aspects of those studies are not yet complete.

Studies were initiated in 1967 to resolve a nagging problem unique to internal emitters—the relative efficiencies for tumor induction of intense, discrete radiation sources and dispersed, low-intensity sources. Specific impetus for these experiments was the atmospheric burn-up in 1964 of a space battery powered by plutonium-238 and a 1974 presentation of a theoretical treatment of "hot particles" as lung carcinogens (Tamplin and Cochran, 1974; discussed in National Council on Radiation Protection and Measurements, 1975, and National Academy of Sciences, 1976). Various intense, discrete alpha-emitting particles and dispersed alpha- and beta-emitting particles were placed in rat and hamster lungs by implantation or intratracheal instillation or through inhalation exposures. On the basis of absorbed dose, distributed low-intensity sources were determined to be more carcinogenic (International Commission on Radiological Protection, 1979). Some aspects of these studies are still in progress.

Experiments were begun in 1971 to examine the distribution and effects in rats and guinea pigs of the major long-lived noble gas fission product, krypton-85. Projections of world use of nuclear energy and fuel reprocessing emphasized the potential importance of this element as a health hazard. Krypton-85 emits only beta particles, which irradiate skin and, to a much lesser degree, lung and fatty tissues (National Council on Radiation Protection and Measurements, 1976). Some late effects of immersion in or inhalation of krypton-85 are still being studied.

Trends in Mammalian Radiobiological Research

The most important trends have been the shifts from descriptive, range-finding studies of radiation effects to examination of quantitative dose-response relationships and from emphasis on acute lethality to studies of late radiation effects. The earliest studies concerned acute effects (particularly lethality) of single, large, high-intensity radiation exposures in small, conventionally reared rodents. Dose-response studies were then expanded to include a variety of end points related to cell damage in specific tissues (e.g., intestinal epithelium, bone marrow, gonads) and to cataloging specific late-appearing diseases (e.g., nephrosclerosis, cancer induction). In addition, the period of observation was extended to the full life span. Large, longer-lived animals were introduced to verify results obtained in rodents, to reveal fundamental species differences in radiation responses, and to improve the ability to extrapolate animal data to humans. Group sizes were increased to improve the statistical reliability of results. New inbred strains were introduced to study the influence of genetics on radiation-induced life-shortening and, specifically, radiation-induced tumors. Specially reared pathogen-free animals were introduced to reduce losses from infectious diseases, thereby maximizing the observation interval and improving the statistical reliability of data on low probability, late-appearing effects.

Trends involving physical radiation factors have shifted toward lower doses and dose rates and toward dose protraction. Photon sources and exposure methods were refined to achieve nearly uniform whole-body irradiation of most experimental animals and to permit prolonged exposures at constant intensity. Neutron dosimetry has been improved, and sources of different neutron energies have been developed to expose adequate numbers of animals to a range of well-characterized neutron doses.

The major trends in research on effects of internally deposited radionuclides have shifted toward the use of longer-lived animals and development of improved dosimetry. Improved methods are being developed for determining nuclide retention in the body, for introducing known amounts of well-characterized radionuclides by inhalation, and for local dosimetry, particularly of bone-seeking radioelements.

REFERENCES

Ainsworth, E. J., R. J. M. Fry, P. C. Brennan, S. P. Stearner, J. H. Rust, and F. S. Williamson. 1976. Life shortening, neoplasia and systematic injuries in mice after single or fractionated doses of neutron or gamma radiation. Pp. 77-92 in Biological and Environmental Effects of Low-Level Radiation. Proceedings of a Symposium on Biological Effects of Low-Level Radiation Pertinent to Protection of Man and His Environment. Organized by the International Atomic Energy Agency and the World Health Organization, held in Chicago, Nov. 3-7, 1975. Vol. I. International Atomic Energy Agency, Vienna.

Andersen, A. C. 1965. The Effects of X-Irradiation on Work Capacity and Longevity of the Dog. Annual Progress Report No. 14. Prepared by the University of California for the U.S. Atomic Energy Commission, San Francisco, Calif.

Andersen, A. C., and L. S. Rosenblatt. 1969. The effect of whole-body x-irradiation on the median lifespan of female dogs. Radiat. Res. 39:177-200.

Andersen, A. C., F. T. Shultz, and T. J. Hage. 1961. The effect of total-body x-irradiation on reproduction of the female beagle to four years of age. Radiat. Res. 15:745-753.

Andersen, A. C., A. G. Hendricks, and M. H. Momeni. 1977. Fractionated x-radiation damage to developing ovaries in the bonnet monkey (*Macaca radiata*). Radiat. Res. 71:398-405.

Bair, W. J. 1970. Plutonium Inhalation Studies. (A series of lectures given in Japan in 1969 at the invitation of the Japanese Atomic Energy Commission.) BNWL-1221. Battelle Pacific Northwest Laboratories, Richland, Wash.

Bair, W. J., J. E. Ballou, J. F. Park, and C. L. Sanders. 1973. Plutonium in soft tissues with emphasis on the respiratory tract. Pp. 503-568 in H. C. Hodge, J. N. Stannard, and J. B. Hursh, eds. Uranium-Plutonium Transplutonic Elements. Springer-Verlag, New York.Blair, H. A., ed. 1954. Biological Effects of External Radiation. National Nuclear Energy Series, Manhattan Project Technical Section, Division VI-Vol. 2. McGraw-Hill Book Co., Inc., New York. 508 pp.

Bloom, W., ed. 1948. Histopathology of Irradiation from External and Internal Sources. National Nuclear Energy Series, Manhattan Project Technical Section, Division IV-

Plutonium Project Record, Vol. 22 I. McGraw-Hill Book Co., Inc., New York. 808 pp.

Bond, V. P. 1969. Radiation mortality in different mammalian species. Pp. 5-19 in V. P. Bond and T. Sugahara, eds. Comparative Cellular and Species Radiosensitivity. Igaku Shoin, Ltd., Tokyo.

Bond, V. P., and T. Sugahara, eds. 1969. Comparative Cellular and Species Radiosensitivity. Igaku Shoin, Ltd., Tokyo. 319 pp.

Bond, V. P., T. M. Fliedner, and J. O. Archambeau. 1965. Mammalian Radiation Lethality. A Disturbance in Cellular Kinetics. Academic Press, New York. 340 pp.

Brown, D. G., R. G. Cragle, and T. R. Noonan, eds. 1968. Proceedings of a Symposium on Dose Rate in Mammalian Radiation Biology, April 29-May 1, 1968, Oak Ridge, Tenn. U.S. Atomic Energy Commission CONF-680410. National Technical Information Service, Springfield, Va.

Brues, A. M., and G. A. Sacher. 1950. Analysis of mammalian radiation injury and lethality. Pp. 441-465 in J. J. Nickson, ed. Symposium on Radiobiology. The Basic Aspects of Radiation Effects on Living Systems. Oberlin College, June 14-18, 1950. John Wiley & Sons, Inc., New York.

Bustad, L. K., ed. 1963. The Biology of Radioiodine. Proceedings of the Hanford Symposium held at Richland, Washington, June 17-19, 1963. Health Phys. 9:1081-1423.

Bustad, L. K., L. A. George, Jr., S. Marks, D. E. Warner, C. M. Barnes, K. E. Herde, and H. A. Kornberg. 1957. Biological effects of I^{131} continuously administered to sheep. Radiat. Res. 6:380-413.

Cantril, S. T. 1951. Biological bases for maximum permissible exposures. Pp. 36-74 in R. S. Stone, ed. Industrial Medicine on the Plutonium Project. Survey and Collected Papers. McGraw-Hill Book Co., New York.

Casarett, G. W., and H. A. Eddy. 1968. Fractionation of dose in radiation-induced male sterility. P. 141 in D. G. Brown, R. G. Cragle, and T. R. Noonan, eds. Proceedings of a Symposium on Dose Rate in Mammalian Radiation Biology, April 29-May 1, 1968, Oak Ridge, Tenn. U.S. Atomic Energy Commission CONF-680410. National Technical Information Service, Springfield, Va.

Clarke, W. J., E. B. Howard, and P. L. Hackett, eds. 1968. Myeloproliferative Disorders of Animals and Man. Proceedings of the Eighth Annual Hanford Biology Symposium at Richland, Washington, May 20-23, 1968. Symposium Series No. 19. U.S. Atomic Energy Commission, Division of Technical Information, Oak Ridge, Tenn. 765 pp.

Coppenger, C. J., and S. O. Brown. 1967. The gross manifestations of continuous gamma irradiation on the prenatal rat. Radiat. Res. 31:230-242.

Cross, F. T. 1978. Study of the Combined Effects of Smoking and Inhalation of Uranium Ore Dust, Radon Daughters, and Diesel Oil Exhaust Fumes in Hamsters and Dogs. PNL-2744, Battelle Pacific Northwest Laboratory, Richland, Wash.

Darden, E. B., Jr. 1969. Current research on factors influencing RBE for mammalian late effects. Pp. 255-267 in Symposium on Neutrons in Radiobiology, Nov. 11-14, 1969, Oak Ridge, Tenn. U.S. Atomic Energy Commission CONF-691106. National Technical Information Service, Springfield, Va.

Davidson, H. O. 1957. Biological Effects of Whole-Body Gamma Radiation on Human Beings. Published for Operations Research Office, Johns Hopkins University. Johns Hopkins Press, Baltimore, Md.

Dougherty, T. F., W. S. S. Jee, C. W. Mays, and B. J. Stover. 1962. Some Aspects of Internal Irradiation. Proceedings of a Symposium held at the Homestead, Heber, Utah, May 8-11, 1961. Pergamon Press, New York.

Drucker H. 1980. Pacific Northwest Laboratory Annual Report for 1979 to the DOE Assistant Secretary for Environment. Part 1. Biomedical Sciences. PNL-3300 PT1. Pacific Northwest Laboratory, Richland, Wash.

Dunning, G. M., and J. A. Hilcken, eds. 1958. Symposium. The Shorter-Term Biological Hazards of a Fallout Field, Washington, D.C., Dec. 12-14, 1956. Sponsored by the Atomic Energy Commission and the Department of Defense. U.S. Government Printing Office, Washington, D.C. 236 pp.

Erickson, B. H. 1967. Effect of gamma radiation on the prepuberal bovine ovary. Radiat. Res. 3l:441-451.

Erickson, B. H., and P. G. Martin. 1976. Effects of continuous prenatal radiation on the pig and rat. Pp. 111-118 in Biological and Environmental Effects of Low-Level Radiation. Proceedings of a Symposium on Biological Effects of Low-Level Radiation Pertinent to Protection of Man and His Environment. Organized by the International Atomic Energy Agency and the World Health Organization, Chicago, Nov. 3-7, 1975. Vol. I. International Atomic Energy Agency, Vienna.

Erickson, B. H., and R. A. Reynolds. 1978. Oogenesis, follicular development and reproductive performance in the prenatally irradiated bovine. Pp. 199-205 in Late Biological Effects of Ionizing Radiation. Proceedings of the Symposium on the Late Biological Effects of Ionizing Radiation, held by the International Atomic Energy Agency in Vienna, March 13-17, 1978. Vol. 2. International Atomic Energy Agency, Vienna.

Erickson, B. H., R. A. Reynolds, and R. L. Murphree. 1976. Late effects of ^{60}Co radiation on the bovine oocyte as reflected by oocyte survival, follicular development, and reproductive performance. Radiat. Res. 68:132-137.

Evans, R. D. 1943. Protection of radium dial workers and radiologists from injury by radium. J. Ind. Hyg. Toxicol. 25: 253-274.

Evans, T. C. 1952. The influence of quantity and quality of radiation on the biologic effect. Pp. 393-413 in J. J. Nickson, ed. Symposium on Radiobiology. The Basic Aspects of Radiation Effects on Living Systems, Oberlin College, June 14-18, 1950. Sponsored by the National Research Council, National Academy of Sciences. John Wiley & Sons, Inc., New York.

Fink, R. M., ed. 1950. Biological Studies with Polonium, Radium, and Plutonium. National Nuclear Energy Series. Manhattan Project Technical Section Division VI-Vol. 3. McGraw-Hill Book Co., Inc., New York. 411 pp.

Finkel, M. P. 1959. Late effects of internally deposited radioisotopes in laboratory Animals. Radiat. Res. Suppl. 1:265-279.

Finkel, M. P., P. J. Bergstrand, and B. O. Biskis. 1960. The consequences of the continuous ingestion of Sr^{90} by mice. Radiology 74:458-467.

Fritz, T. E., W. P. Norris, C. M. Poole, and C. E. Rehfeld. 1966. Cerium-144—Toxic manifestations in beagle dogs. Pp. 112-113 in ANL-7278. Argonne National Laboratory, Argonne, Ill.

Fritz, T. E., W. P. Norris, D. V. Tolle, T. M. Seed, C. M. Poole, L. S. Lombard, and D. E. Doyle. 1978. Relationship of dose rate and total dose to responses of continuously irradiated beagles. Pp. 71-82 in Late Biological Effects of Ionizing Radiation. Proceedings of the Symposium on the Late Biological Effects of Ionizing Radiation held by the International Atomic Energy Agency, Vienna, March 13-17, 1978. Vol. 2. International Atomic Energy Agency, Vienna.

Furth, J., and E. Lorenz. 1954. Carcinogenesis by ionizing radiations. Pp. 1145-120l in A. Hollaender, ed. Radiation Biology. Vol. 1: High Energy Radiation. McGraw-Hill Book Co., Inc., New York.

Goldman, M., R. J. Della Rosa, and M. H. Momeni. 1972. Radiation dose to beagles from continuous ^{90}Sm exposure. Pp. 103-112 in M. Goldman and L. K. Bustad, eds. Biomedical Implications of Radiostrontium Exposure. Proceedings of a symposium held at Davis, Calif., Feb. 22-24, 1971. AEC Symposium Series 25. U.S. Atomic Energy Commission, Oak Ridge, Tenn.

Grahn, D., and G. A. Sacher. 1968. Fractionation and protraction factors and the late effects of radiation in small mammals. Pp. 2.1-2.27 in D. G. Brown, R. G. Cragle, and T. R. Noonan, eds. Proceedings of a Symposium on Dose Rate in Mammalian Radiation Biology, April 29-May 1, 1968, Oak Ridge, Tenn. U.S. Atomic Energy Commission CONF 680410. National Technical Information Service, Springfield, Va.

Grahn, D., G. A. Sacher, R. A. Lea, R. J. M. Fry, and J. H. Rust. 1978. Analytical approaches to and interpretations of data on time, rate and cause of death of mice exposed to external gamma irradiation. Pp. 43-58 in Late Biological Effects of Ionizing Radiation. Proceedings of the Symposium on The Late Biological Effects of Ionizing Radiation held by the International Atomic Energy Agency in Vienna, March 13-17, 1978. Vol. 2. International Atomic Energy Agency, Vienna.

Hamilton, J. G. 1947. The metabolism of the fission products and the heaviest elements. Radiology 49:269-365.

Henderson, R. F., J. H. Diel, and B. S. Martinez, eds. 1979. Annual Report of the Inhalation Toxicology Research Institute, operated for the U.S. Department of Energy by the Lovelace Biomedical and Environmental Research Institute, Inc. Oct. 1, 1978 through Sept. 30, 1979.

Henshaw, P. S., ed. 1965. Late Irradiation Effects Conference: Inventory of Capability. Summary of Proceedings, Dec. 4-6, 1963. U.S. Atomic Energy Commission, Germantown, Md.

Henshaw, P. S., E. F. Riley, and G. E. Stapleton. 1947. The biologic effects of pile radiations. Radiology 49:349-360.

Holaday, D. A. 1973. Mining hazards. Pp. 295-306 in H. C. Hodge, J. N. Stannard, and J. B. Hursh, eds. Uranium-Plutonium Transplutonic Elements. Springer-Verlag, New York.

Hughes, W. L., V. P. Bond, G. Brecher, E. P. Cronkite, R. B. Painter, H. Quastler, and F. G. Sherman. 1958. Cellular proliferation in the mouse as revealed by autoradiography with tritiated thymidine. Proc. Natl. Acad. Sci. USA 44:476-483.

Hupp, E. W.. 1976. Effects of continuous lifetime irradiation on Spanish goats. Pp. 119-126 in Biological and Environmental Effects of Low-Level Radiation. Proceedings of a Symposium on Biological Effects of Low-Level Radiation Pertinent to Protection of Man and His Environment organized by the International Atomic Energy Agency and the World Health Organization, Chicago, Nov. 3-7, 1975. Vol. I. International Atomic Energy Agency, Vienna.

Hupp, E. W., S. O. Brown, and J. W. Austin. 1965. Continuous and acute irradiation effects on developing animals. Pp. 589-600 in M. R. Sikov and D. D. Mahlum, eds. Radiation Biology of the Fetal and Juvenile Mammal. Proceedings of the Ninth Annual Hanford Biology Symposium at Richland, Wash., May 5-8, 1969. AEC Symposium Series 17. U.S. Atomic Energy Commission, Oak Ridge, Tenn.

International Commission on Radiological Protection. 1959. Report of Committee II on Permissible Dose for Internal Radiation (1959), with Bibliography for Biological, Mathematical and Physical Data. Health Physics, Vol. 3. Pergamon Press, New York. 233 pp.International Commission on Radiological Protection. 1979. Biological Effects of Inhaled Radionuclides. Radiation Protection. ICRP Publ. 31. Pergamon Press, New York.

Law, L. W. 1960. Radiation carcinogenesis. Adv. Biol. Med. Phys. 7:295-342.

Leach, L. J., E. A. Maynard, H. C. Hodge, J. K. Scott, C. L. Yuile, G. E. Sylvester, and H. B. Wilson. 1970. A five-year inhalation study with natural uranium dioxide (UO_2) dust. I. Retention and biologic effect in the monkey, dog, and rat. Health Phys. 18:599-612.

Leach L. J., C. L. Yuile, H. C. Hodge, G. E. Sylvester, and H. B. Wilson. 1973. A five-year inhalation study with natural uranium dioxide (UO_2) dust. II. Postexposure retention and biologic effects in the monkey, dog, and rat. Health Phys. 5:239-258.

Lesher, S. 1966. Chronic irradiation and ageing in mice and rats. Pp. 183-206 in P. J. Lindop and G. A. Sacher, eds. Radiation and Ageing. Proceedings of a Colloquium held in Semmering, Austria, June 23-24, 1966. Taylor and Francis, Ltd., London.

Lisco, H., M. P. Finkel, and A. M Brues. 1947. Carcinogenic properties of radioactive fission products and of plutonium. Radiology 49:361-363.

Lorenz, E., W. E. Heston, A. B. Eschenbrenner, and M. K. Deringer. 1947. Biological studies in the tolerance range. Radiology 49:274-285.

Lorenz, E., L. O. Jacobson, W. E. Heston, M. Shimkin, A. B. Eschenbrenner, M. K. Deringer, J. Doniger, and R. Schweisthal. 1954. Effects of long-continued total-body gamma irradiation on mice, guinea pigs, and rabbits. III. Effects on life span, weight, blood picture, and carcinogenesis and the role of the intensity of radiation. Pp. 24-148 in R. E. Zirkle, ed. Biological Effects of External X and Gamma Radiation. Part I. McGraw Hill Book Co., Inc., New York.

Mays, C. W. 1978. Biological Effects of ^{224}Ra and Thorotrast. Proceedings of an International Symposium, held at Alta, Utah, July 21-23, 1974. Health Phys. 35:3-174.

Mays, C. W., and M. P. Finkel. 1980. RBE of α-particles vs. β-particles in bone sarcoma induction. Pp. 401-405 in Proceedings of the Fifth International Congress of IRPA, Jerusalem, Israel, Vol. II.

Mays, C. W., W. S. S. Jee, R. D. Lloyd, B. J. Stover, J. H. Dougherty, and G. N. Taylor. 1969. Delayed Effects of Bone-Seeking Radionuclides. Sun Valley, Sept. 12-14, 1967. University of Utah Press, Salt Lake City.

Mutscheller, A. 1928. Safety standards of protection against x-ray dangers. Radiology 10:468-476.

National Academy of Sciences. 1956. Report of the Committee on Pathologic Effects of Atomic Radiation. From a Study of the Biological Effects of Atomic Radiation. Publ. 452. National Academy of Sciences, Washington, D.C. 200 pp.

National Academy of Sciences. 1961. Long-Term Effects of Ionizing Radiations from External Sources. Committee on Pathologic Effects of Atomic Radiation. Publ. 849. National Academy of Sciences, Washington, D.C. 82 pp.

National Academy of Sciences. 1976. Health Effects of Alpha-Emitting Particles in the Respiratory Tract. Report of Ad Hoc Committee on "Hot Particles" of the Advisory Committee on the Biological Effects of Ionizing Radiations. National Academy of Sciences, Washington, D.C. 107 pp.

National Academy of Sciences. 1980. The Effects on Populations of Exposure to Low Levels of Ionizing Radiation. Report of the Committee on the Biological Effects of Ionizing Radiations. National Academy of Sciences, Washington, D.C. 638 pp.

National Bureau of Standards. 1941. Handbook H27. Safe Handling of Radioactive Luminous Compound. U.S. Government Printing Office, Washington, D.C.

National Council on Radiation Protection and Measurements. 1959. Maximum Permissible Body Burdens and Maximum Permissible Concentrations of Radionuclides in Air and in Water for Occupational Exposures. NCRP Report No. 22. National Council on Radiation Protection and Measurements, Washington, D.C.

National Council on Radiation Protection and Measurements. 1975. Alpha-Emitting

Particles in Lungs. NCRP Report No. 46. National Council on Radiation Protection and Measurements, Washington, D.C. 28 pp.

National Council on Radiation Protection and Measurements. 1976. Krypton-85 in the Atmosphere—Accumulation, Biological Significance, and Control Technology. Recommendations of the NCRP. NCRP Report No. 44. National Council on Radiation Protection and Measurements, Washington, D.C.

National Council on Radiation Protection and Measurements. 1977. Cesium-137 from the Environment to Man: Metabolism and Dose. Recommendations of the NCRP. NCRP Report. No. 52. National Council on Radiation Protection and Measurements, Washington, D.C.

National Council on Radiation Protection and Measurements. 1980. Influence of Dose and Its Distribution in Time on Dose-Response Relationships for Low-LET Radiations. Recommendations of the NCRP. NCRP Report. 64. National Council on Radiation Protection and Measurements, Washington, D.C.

Norris, W. P., and T. E. Fritz. 1972. Interactions of total dose and dose rate in determining tissue responses to ionizing radiations. Pp. 243-260 in B. J. Stover and W. S. S. Jee, eds. Radiobiology of Plutonium. The J. W. Press, University of Utah, Salt Lake City.

Norris, W. P., and T. E. Fritz. 1974. Response of the beagle dog to protracted exposure to ^{60}Co gamma rays. Radiat. Res. 59:145. (abstract)

Norris, W. P., S. A. Tyler, and G. A. Sacher. 1976. An interspecies comparison of responses of mice and dogs to continuous ^{60}Co irradiation. Pp. 147-150 in Biological and Environmental Effects of Low-Level Radiation. Proceedings of a Symposium on Biological Effects of Low-Level Radiation Pertinent to Protection of Man and His Environment organized by the International Atomic Energy Agency and the World Health Organization, Chicago, Nov. 3-7, 1975. Vol. I. International Atomic Energy Agency, Vienna.

Oakberg, E. F. 1955. Sensitivity and time of degeneration of spermatogenic cells irradiated in various stages of maturation in the mouse. Radiat. Res. 2:369-391.

Oakberg, E. F. 1966. Effect of 25 R of x-rays at 10 days of age on oocyte numbers and fertility of female mice. Pp. 293-306 in P. J. Lindop and G. A. Sacher, eds. Radiation and Ageing. Proceedings of a Colloquium held in Semmering, Austria, June 23-24, 1966. Taylor and Francis, Ltd., London.

Oakberg, E. F. 1968. Mammalian gametogenesis and species comparisons in radiation response of the gonads. Pp. 3-15 in Effects of Radiation on Meiotic Systems. Report of a Study Group on the Effects of Radiation on Meiotic Systems, organized by the International Atomic Energy Agency, held in Vienna, May 8-11, 1967. International Atomic Energy Agency, Vienna.

Patt, H. M., and A. M. Brues. 1954a. The pathological physiology of radiation injury in the mammal. I. Physical and biological factors in radiation action. Pp. 919-958 in A. Hollaender, ed. Radiation Biology. Vol. I: High Energy Radiation. McGraw-Hill Book Co., Inc., New York.

Patt, H. M. and A. M. Brues. 1954b. The pathological physiology of radiation injury in the mammal. II. Specific aspects of the physiology of radiation injury. Pp. 959-1028 in A. Hollaender, ed. Radiation Biology. Vol. 1: High Energy Radiation. McGraw-Hill Book Co., Inc., New York.

Phemister, R. D., R. J. Garner, J. N. Shively and W. J. Tietz. 1969. Radiosensitivity of the developing beagle. Pp. 395-406 in M. R. Sikov and D. D. Mahlum, eds.

Radiation Biology of the Fetal and Juvenile Mammal. Proceedings of the Ninth Annual Hanford Biology Symposium, Richland, Wash., May 5-8, 1969. AEC Symposium Series 17. U.S. Atomic Energy Commission, Oak Ridge, Tenn.

Prosser, C. L., E. E. Painter, H. Lisco, A. M. Brues, L. O. Jacobson, and M. N. Swift. 1947. The clinical sequence of physiological effects of ionizing radiation in animals. Radiology 49:299-313.

Raabe, O. G., H. A. Boyd, G. M. Kanapilly, C. J. Wilkinson, and G. J. Newton. 1975. Development and use of a system for routine production of monodisperse particles of $^{238}PuO_2$ and evaluation of gamma emitting labels. Health. Phys. 28:655-667.

Russell, L. B. 1954. The effects of radiation on mammalian prenatal development. Pp. 861-918 in A. Hollaender, ed. Radiation Biology. Vol. I: High Energy Radiation. McGraw-Hill Book Co., Inc., New York.

Sacher, G. A. 1958. Reparable and irreparable injury: A survey of the position in experiment and theory. Pp. 283-313 in W. D. Claus, ed. Radiation Biology and Medicine. Addison-Wesley Publishing Co., Reading, Mass.

Sacher, G. A., and D. Grahn. 1964. Survival of mice under duration-of-life exposure to gamma rays. I. The dosage-survival relation and the lethality function. J. Natl. Cancer Inst. 32:277-321.

Sanders, C. L., Jr., R. C. Thompson, and W. J. Bair. 1970. Lung cancer: Dose response studies with radionuclides. Pp. 285-304 in M. G. Hanna, Jr., P. Nettesheim, and J. R. Gilbert, eds. Inhalation Carcinogenesis. Proceedings of a Biology Division, Oak Ridge National Laboratory, conference held in Gatlinburg, Tenn., Oct. 8-11, 1969. AEC Symposium Series 18. U.S. Atomic Energy Commission, Oak Ridge, Tenn.

Sanders, C. L., R. H. Busch, J. E. Ballou, and D. D. Mahlum, eds. 1973. Radionuclide Carcinogenesis. Proceedings of the Twelfth Annual Hanford Biology Symposium at Richland, Wash., May 10-12, 1972. Atomic Energy Commission Symposium Series #29, CONF-720505. National Technical Information Service, Springfield, Va. 500 pp.

Schambra, P. E., G. E. Stapleton, and N. F. Barr. 1967. Space Radiation Biology. Proceedings of a Workshop Conference on Space Radiation Biology sponsored by the National Aeronautics and Space Administration held at the University of California, Berkeley, Calif., Sept 7-10, 1965. Radiat. Res. Suppl. 7.

Scott, K. G., D. Axelrod, J. Crowley, and J. G. Hamilton. 1949. Deposition and fate of plutonium, uranium and their fission products inhaled as aerosols by rats and man. Arch. Pathol. 48:31-54.

Sievert, R. M. 1925. Einig Untersuchungen ueber Vorrichtungen zum Schutzgegen Roentgenstrahlen. Acta Radiol. 4:61-75.

Sikov, M. R., and D. D. Mahlum, eds. 1969. Radiation Biology of the Fetal and Juvenile Mammal. Proceedings of the Ninth Annual Hanford Biology Symposium, Richland, Wash., May 5-8, 1969. Atomic Energy Commission Symposium Series #17. CONF-69051 National Technical Information Service, Springfield, Va. 1,026 pp.

Siri, W. E., E. C. Dougherty, C. A. Tobias, R. W. Dunn, J. S. Robertson, and P. P. Weymouth. 1949. Isotopic Tracers and Nuclear Radiations with Applications to Biology and Medicine. McGraw-Hill Book Co., Inc., New York.

Spaulding, J. F., J. R. Prine, and G. L. Tietjen. 1978. Late biological effects of ionizing radiation as influenced by dose, dose rate, age at exposure and genetic sensitivity to neoplastic transformation. Pp. 3-11 in Late Biological Effects of Ionizing radiation. Vol. 2. Proceedings of the Symposium on the Late Biological Effects of Ionizing

Radiation, held by the International Atomic Energy Agency, Vienna, March 13-17, 1978. International Atomic Energy Agency, Vienna.

Stannard, J. N. 1973. Plutonium in the environment. Pp. 669-688 in H. C. Hodge, J. N. Stannard, and J. B. Hursh, eds. Uranium-Plutonium Transplutonic Elements. Springer-Verlag, New York.

Stannard, J. N., and G. W. Casarett. 1964. Metabolism and biological effects of an alpha particle emitter, polonium-210. Radiat. Res. Suppl. 5.1-437.

Stone, R. S., ed. 1951. Industrial Medicine on the Plutonium Project. Survey and Collected Papers. National Nuclear Energy Series. Manhattan Project Technical Section. Division IV—Plutonium Project Record. Vol 20. McGraw-Hill Book Co., Inc., New York.

Storer, J. B., and P. C. Sanders. 1958. Relative effectiveness of neutrons for production of delayed biological effects. I. Effect of single doses of thermal neutrons on life span of mice. Radiat. Res. 8:64-70.

Storer, J. B., P. S. Harris, J. E. Furchner, and W. H. Langham. 1957. The relative biological effectiveness of various ionizing radiations in mammalian systems. Radiat. Res. 6:188-288.

Storer, J. B., B. S. Rogers, I. U. Boone, and P. S. Harris. 1958. Relative effectiveness of neutrons for production of delayed biological effects. II. Effect of single doses of neutrons from an atomic weapon on life span mice. Radiat. Res. 8:71-76.

Storer, J. B., L. J. Serrano, E. B. Darden, Jr., M. C. Jernigan, and R. L. Ullrich. 1979. Life shortening in RFM and BALB/c mice as a function of radiation quality, dose, and dose rate. Radiat. Res. 78:122-161.

Stover, B. J., W. S. S. Jee, and J. S. Montague, eds. 1972. Radiobiology of Plutonium. J. W. Press, University of Utah, Salt Lake City.

Tamplin, A. R., and T. B. Cochran. 1974. A Report on the Inadequacy of Existing Radiation Protection Standards Related to Internal Exposures of Man to Insoluble Particles of Plutonium and Other Alpha Emitting Hot Particles. Natural Resources Defense Council, Washington, D.C.

Taylor, J. H. 1960. Autoradiography with tritium-labeled substances. Med. Phys. 7:107-130.

Thomassen, R. W., G. M. Angleton, A. C. Lee, R. D. Phemister, and A. Benjamin. 1978. Neoplasms in dogs receiving low-level gamma radiation during pre- and postnatal development. Pp. 181-189 in Late Biological Effects of Ionizing Radiation. Proceedings of the Symposium on the Late Biological Effects of Ionizing Radiation held by the International Atomic Energy Agency, Vienna, March 13-17, 1978. Vol. 2. International Atomic Energy Agency, Vienna.

Tobias, C. A. 1950. The dependence of some biological effects of radiation on the rate of energy loss. Pp. 357-392 in J. J. Nickson, ed. Symposium on Radiobiology. The Basic Aspects of Radiation Effects on Living Systems. Oberlin College, June 14-18, 1950. John Wiley & Sons, Inc., New York.

Ullrich, R. L., and J. B. Storer. 1979a. Influence of γ irradiation on the development of neoplastic disease in mice. I. Reticular tissue tumors. Radiat. Res. 80:303-316.

Ullrich, R. L., and J. B. Storer. 1979b. Influence of γ irradiation on the development of neoplastic disease in mice. II. Solid tumors. Radiat. Res. 80:317-324.

Ullrich, R. L., and J. B. Storer. 1979c. Influence of γ irradiation on the development of neoplastic disease in mice. III. Dose-Rate Effects. Radiat. Res. 80:325-342.

Upton, A. C., A. W. Kimball, J. Furth, K. W. Christenberry, and W. H. Benedict. 1960. Some delayed effects of atom-bomb radiations in mice. Cancer Res. 20:1-62.

Upton, A. C., R. C. Allen, R. C. Brown, N. K. Clapp, J. W. Conklin, G. E. Cosgrove,

E. B. Darden, Jr., M. A. Kastenbaum, T. T. Odell, Jr., L. J. Serrano, R. L. Tyndall, and H. E. Walburg, Jr. 1969. Quantitative experimental study of low-level radiation carcinogenesis. Pp. 425-438 in Radiation-Induced Cancer. Proceedings of a Symposium on Radiation-Induced Cancer, organized by the International Atomic Energy Agency and the World Health Organization, Athens, April 28-May 2, 1969. International Atomic Energy Agency, Vienna.

Van Cleave, C. D. 1968. Late Somatic Effects of Ionizing Radiation. U.S. Atomic Energy Commission, Oak Ridge, Tenn. National Technical Information Service, Springfield, Va. 310 pp.

Voegtlin, V., and H. C. Hodge, eds. 1949 and 1953. Pharmacology and Toxicology of Uranium Compounds. National Nuclear Energy Series, Division VI, Vol. 1. Parts 1 and 2 (1949), Parts 3 and 4 (1953). McGraw-Hill Book Co., Inc., New York.

Walburg, H. E., Jr. 1975. Radiation-induced life-shortening and premature aging. Adv. Radiat. Biol. 5:147-179.

Wrenn, M. E. 1979. Research in Radiobiology. Annual Report of Work in Progress in the Internal Irradiation Program. Radiobiology Division, Department of Pharmacology, University of Utah College of Medicine, Salt Lake City. COO-119-254.

Zirkle, R. E., ed., 1951. Effects of External Beta Radiation. McGraw-Hill Book Co., Inc., New York.

Zirkle, R. E., ed. 1954. Biological Effects of External X and Gamma Radiation. Part 1. McGraw-Hill Book Co., Inc., New York. 530 pp.

Zirkle, R. E., ed. 1956. Biological Effects of External X and Gamma Radiation. Part 2. U.S. Atomic Energy Commission. National Technical Information Service, Springfield, Va.

6

Biological Effects of Ionizing Radiation

Studies of systems other than whole human beings are necessary to understand the effects of radiation and, in turn, to gain an understanding of the prediction of effects in whole organisms and the prevention of injury. This chapter contains discussions of the cellular, subcellular, and molecular effects of radiation and the repair processes that are activated upon injury. The total effect at the cellular level is viewed as the sum of the induced injuries (lesions) plus the effect of repair mechanisms. Related special studies of radiation effects in animals and in ecosystems are discussed in other chapters.

At the cellular and subcellular levels of study *in vitro*, the "control" systems found in whole organisms, such as the endocrine, immunologic, and inflammatory systems, are generally inoperative. The absence of these modulating forces permits the direct interpretation of the cellular findings since cellular effects are the alterations from which all body changes originate. The interaction of radiation science and basic biology had its earliest major impact in 1927, when Muller discovered that radiation could "transmutate" the gene (Muller, 1927, 1928). This observation and the subsequent discovery that many of the mutations were associated with aberrant chromosomes helped to revitalize the fields of genetics and cytogenetics. Radiation has been used to produce a multitude of gene and chromosome mutations that permitted the investigation of such biological phenomena as mutagenesis, gene expression, mutagenic loads, and chromosome kinetics.

The development of an understanding of the kinetics of the induction of genetic damage led to the formulation of the target theory (Lea, 1946), in which the effects of radiation are attributed to a direct interaction of radiant energy with specific target sites, or molecules, within the cell. The mathematical formulations derived from the target theory studies have had a significant impact on our understanding of cell growth, especially under *in vitro* culture conditions, and cell kinetics. The association of radiosensitivity and cell proliferation phenomena gave impetus to the development of radioisotope-labeled precursors of DNA (e.g., tritiated thymidine), which, in turn, has led to present-day concepts of the proliferative cell cycle. It was also fundamental to new approaches that permitted studies of the origin of functional cells in a variety of cell renewal systems. Through these developments, the relationship of the cell cycle and DNA synthesis was elaborated. The ability to induce and recognize lesions in the genetic material of the cell also led to the complementary study of the repair of such lesions.

The uses of radiobiologic theory and the tools that it provided are many:

- The field of immunology was restimulated by the observation that lethally irradiated animals could survive if their spleens were shielded from exposure or if they were injected with unirradiated bone marrow (Jacobson *et al.*, 1949; Lorenz *et al.*, 1952). The production of chimeric animals by irradiation and bone marrow replacement from other species and genera opened the field of transplantation immunology and the study of graft-versus-host reactions.
- In plant breeding, radiation is used as a source of new mutations to produce additional genetic variablity and new characteristics that are useful for humans.
- Release of radiation-sterilized male insect pests to compete with the unsterilized wild population has served as a biological control system to reduce destructive insect populations.
- Analysis of the effects of radiation on cells has led to a deeper understanding of the effects of chemical mutagens and carcinogens. A number of chemical agents seem to mimic ionizing radiation damage in one of several ways (Setlow, 1978).
- The methods for estimating dose-response relationships first used to evaluate genetic and somatic risks from radiation have contributed to toxicological theory and are being used to estimate the risk arising from the exposure of humans to environmental chemicals.

This limited list of examples of the impact of radiation studies on general biology are intended to illustrate the extensive use to which the methods, tools, and theory of radiobiology have been put.

In like manner, understanding of the effects of radiation in the cell, the manner in which the effects are produced, and the manner in which lesions are repaired is dependent upon our knowledge of basic biology. An understanding of radiation genetics requires knowledge of mutagenesis at the molecular, chromosomal, cellular, and body level, as well as knowledge of DNA structure, chromosomal organization, cell division and survival, gametogenesis, and mechanisms of DNA repair. A similar set of biological parameters could be listed for each of the specific radiation effects noted in Chapter 4.

In order to answer the questions about radiation that concern the federal government, the general public, and the scientific community, there is a need for investigations into basic biological mechanisms, whether or not radiation is used as a probe of the system, and investigations of a more programmatic nature designed to measure the degree and character of the biological perturbations brought about by ionizing radiation.

Despite extensive accumulation of information about ionizing radiation and its effects, which may exceed the amount of knowledge that we have concerning any other noxious environmental agent, we still do not understand, nor have the data to evaluate, the basic mechanisms of radiation-induced damage. The ensuing paragraphs discuss fundamental questions that should be answered in order to gain a reasonably comprehensive understanding of radiation effects that would not only have scientific merit but also heuristic value in the resolution of federal regulatory and legal issues. For each question, there is a short discussion of what is known and what remains to be studied. Occasionally, the evidence and theoretical postulates covered in these discussions are highly technical, but the importance of the conceptual material in establishing a basis for this study requires attention to these details.

1. *How can one determine the response to radiation exposure at low doses?*

Reference to the "low-dose question" as it relates to observations in humans (Chapter 4) sets the stage for the subset of questions that can be addressed to cellular and molecular biology. As a necessary introduction to these questions, we must ask the following questions.

2. *What animal studies would be necessary and useful in determining the*

effects of low doses on humans? To what extent can animal studies be extrapolated to humans?

In order to obtain reliable information of low-dose effects in humans, there is need to perform experiments on animals. The ideal experimental animal, with regard to statistical significance, is one with high sensitivity to the radiation-induced effect, combined with low natural incidence. The beagle dog may be the animal of choice for studies of bone cancer. This animal has a very low incidence of bone sarcoma (less than 1%) when unirradiated, but its skeleton seems 25 times more radiosensitive than that of humans (Mays, 1976).

The application of results from animal studies to predictions for humans should be based on several different species of animals when possible. The beagle and other species with similar qualities can be used as a basis for attempts to compare the effectiveness of different types of radiation at low doses.

Animal studies have frequently been directed toward higher doses of radiation, requiring fewer animals to obtain the necessary observations, and have involved such species as Sprague-Dawley rats, which have a much higher sensitivity to radiation-induced cancer than does the human population (Shellabarger *et al.*, 1980). Exposure of the breast tissue to radiation lowers the age at which the tumors are observed in Sprague-Dawley rats, and irradiation of tissues other than the breast has little effect on acceleration of the appearance of the breast tumors (Bond *et al.*, 1960a,b). Despite the statistical reliability of the results of these studies, it is difficult to extrapolate this information to other species. Nevertheless, to make use of these studies, there is a need to develop a theoretical construct whereby animal observations can be reinterpreted for humans.

From experiments conducted with dogs, rats, and mice at higher levels of radiation or with highly sensitive strains of these animals at lower doses, it can be concluded with confidence that there are situations in which dose-response curves are not linear (Mays and Lloyd, 1972; National Council on Radiation Protection and Measurements, 1980). On the other hand, data for some tumors suggest that the dose-response curve is approximately linear even for high-energy x-rays (Bond *et al.*, 1960a,b). In most cases reported to date, the dose-response curves for high-LET radiation, i.e., heavily charged particles of the kind that would be produced by neutron irradiation, have been approximately linear. Frequently, the information required for extrapolation of data from these experiments to effects in humans is not available, and there are few means of verifying any proposed extrapolation with direct results in humans.

Since there are no adequate data for humans for many types of exposure to ionizing radiation, e.g., inhalation deposition of plutonium-239 in the lung or deposition of plutonium-239 on bone surfaces, experiments in animals that can provide toxicity ratios should be considered. These studies of toxicities of radionuclides are designed to translate the known risk to humans under one condition into a reasonable prediction of risk to humans under another condition. For example, experiments in beagle dogs injected with plutonium-239 or radium-226 indicate that, in terms of the average skeletal dose in rads, bone-surface-seeking plutonium-239 is approximately 16 times more toxic than bone-volume-seeking radium-226 in the induction of bone sarcomas (Mays and Finkel, 1980). Since we have an estimation of the risk of radium-226 in humans from studies of radium dial painters, a useful assumption would be that plutonium-239 would also be approximately 16 times as damaging in humans. Similar use of toxicity ratios from animal studies has led the International Commission on Radiological Protection (1977) to conclude that the quality factor for alpha particles is 20, meaning that the permissible dose in rads from alpha particles is 20 times less than that for x-rays or gamma rays.

3. *Why do different species exhibit differences in radiation sensitivity?*

If this question could be answered unequivocally, much of the problem of extrapolation from animal to humans could be attacked by examining the factors in each system that determine susceptibility or resistance.

The range of sensitivity among living species runs from bacteria that can withstand doses of several thousand rads to mammalian cells and whole mammals that can be killed by only a few hundred rads (Hollaender, 1954). This difference in lethality, as an end point, also exists for the induction of mutations (Abrahamson *et al.*, 1973). In higher forms of life, cell lethality has been correlated with the induction of chromosomal aberrations and breaks. These alterations can lead to the loss or inactivation of vital genes. This type of killing is dependent upon the organization of DNA into chromosomes and the efficiency of whatever repair systems may be available. In general, as one goes up the phylogenetic scale from bacteria to fungi to insects to mammals, the amount of DNA increases, as does sensitivity to radiation (Abrahamson *et al.*, 1973). It has been suggested that the sensitivity of the organized genetic material is related to increases in the size or structural complexity of the mutable target (Abrahamson *et al.*, 1973). Since it is likely that other factors can modify the responses to radiation injury, it is important that experiments directed toward identifying such factors be undertaken.

Less well understood are the differences that are found *within* a species. For example, the RF strain of mice has been found to be more sensitive for leukemia induction by irradiation than are other strains of the same species (Upton *et al.*, 1958). Obviously, genetic factors are behind such differences in sensitivity, but the basic mechanisms responsible for the differences are unknown.

Similar differences in radiation sensitivity have been observed in humans. Some patients with head and neck cancers undergoing radiotherapy have been found to react excessively to normal therapeutic regimes. Furthermore, cells from humans with Down's syndrome (extra chromosomes), other trisomies, ataxia telangiectasia (associated with a deficiency in immune response and defective DNA repair), and Fanconi's anemia differ from normal cells in their susceptibility to radiation injury. In the absence of a coherent theory of radiation biology, the species and intraspecies variations in sensitivity to radiation make the problems of direct extrapolation from animals to humans unusually difficult.

In addition to the differences just described, when various organisms are exposed to internal emitters, they exhibit different sensitivities that are associated with a number of additional factors. The impact of internal emitters is in part determined by their biological half-lives. Physiological factors, such as cell turnover rates, organ size, body mass, and immune surveillance, influence the number and life-span of the target cells and the distribution of dose. Thorough understanding of the factors that influence sensitivity to radiation would be a major step toward solving the problems of extrapolation from animals to humans.

4. *Can studies at the cellular and subcellular level be of value in determining the effects of low doses of radiation on whole organisms, including humans?*

An alternative to direct human or whole animal studies to obtain information about the true shape of the dose-response curves for carcinogenesis and different types of studies for other types of radiation damage is reliance on an increased understanding of the basic mechanisms of carcinogenesis. The study of systems best observed by experiments with cells and with the isolated smaller component systems of subcellular and molecular biology should be productive.

Such studies could clarify not only the mechanisms and immediate target effects of ionizing radiation, but also help to delineate the factors that enter into the final product, a malignant tumor. The questions of linearity or nonlinearity of response to exposure could then be derived from a basic understanding of the mechanisms of alteration and repair.

These investigations can and should be conducted with whatever organism and at whatever level of systematic complexity is most suitable for the question and would include bacteria, plants, and lower animals as well as isolated cells, cells in culture, and subcellular or molecular systems. For example, dose-response curves for the production of pink mutations in the spiderwort *Tradescantia* resulting from acute radiation doses have been carried down to 0.25 rad for x-rays and to 0.01 rad for neutrons. At these low doses, the number of radiation-induced events is only a fraction of the background level, but can be detected with reliability because the affected tissues, stamen hairs, are present in tremendous numbers.

The first question one might ask about the nonlinear dose-response curves suggested by data from animal experiments is whether they are due to an initial transformation of a single cell or more than one cell. It is difficult at this time to suggest experiments that might be used to test the second possibility. Although there is *post hoc* evidence that a tumor can be derived from a single transformed cell, this does not rule out the initial generation of multiple clones from several transformed cells, one of which outgrew the rest. It could be that several transformed cells in the same neighborhood are needed before any one of them could reproduce to form a tumor. If this should prove to be the case, additional questions could be posed about the role of local cell-to-cell communication and control.

The hypothesis that nonlinearity is a function of the response of individual cells is simpler to test. Although there are pitfalls with experiments in which tissue culture cells are irradiated and tested to determine whether tumor transformations have taken place, evidence suggests that some cell systems show a nonlinear response (Kellerer and Rossi, 1972), and several of them have been analyzed by rather sophisticated mathematical models (Kellerer and Rossi, 1972). A specific mathematical model generated from these observations suggests that radiation-induced lesions occur close to each other in time and space (Kellerer and Rossi, 1972; Lea, 1946). However, the molecular nature of such postulated lesions is not known, although it has been repeatedly postulated that chromosomal aberrations could be responsible (Cairns, 1981). Development of investigations appropriate to identifying such postulated lesions are essential.

Since expression of the effects of radiation injury is due to a summation of lesions and their repair, attention must be directed toward the repair mechanisms. Studies with the bacterium *Escherichia coli* have demonstrated that the proteins responsible for repair of certain types of chemical damage to DNA are themselves induced (Samson and Cairns, 1977). Furthermore, the repair proteins are "used up" during the repair process, so that systems can cope only with the correction of a limited number of damaged sites. The sum of damage and repair results in a very nonlinear system that can correct damage at a low dose but not that occurring at a higher dose. A similar phenomenon has recently been reported for chemical damage to mammalian cells (Samson and Schwartz, 1980).

Repair systems with these properties have not been found for ionizing radiation damage. If they do exist, they could offer one explanation for the nonlinear portions of the dose-response curves found at low doses of x-rays. In principle, it would then be possible to evaluate and predict the shape of the dose-response curve for humans by measuring the repair proteins in the cells of an individual after various x-ray exposures. Such tests could be conducted simply by irradiating a few cells removed from the organism. This indicates one direction in which cellular and molecular research might proceed to attempt to gain a powerful method for the evaluation of the low-dose risk for humans or for other species.

Other molecular, cellular, and genetic processes play determinant roles in the response to injury and may account, collectively or individually, for the shapes of dose-response curves. A number of assumptions can be made by using a form of radiation "target theory" to account for almost any observed dose-response curve (Lea, 1946). However, such theoretical treatments do not have much predictive value for radiation-induced cancers at low doses until one is able to identify the molecular entities that constitute the "targets." Once such an identification can be made, however, one would have a method for evaluating the theory as well as a way of testing individuals to determine their probable response to radiation.

Thus far in this discussion, primary attention has been directed toward the problem of radiation carcinogenesis as the measured effective end point. Genetic effects of low-dose radiation require similar attention.

The genetic effects of radiation can be the result of gene mutations. These can be more discretely specified as gene mutations caused by effects on DNA, such as base changes, frameshifts, and small deletions, or gross chromosomal mutations, such as large deletions, translocations, inversions, and aneuploidy.

Recent estimates of the genetic risk from radiation (National Academy of Sciences, 1980) have involved calculating the probable relative increase in the number of human genetic disorders (and genetically determined ill-health) brought about by a dose of 1 rem. The BEIR III Committee used estimates of the incidence of various genetic disorders in humans and estimates of radiation-induced mutations in mice, both of which are uncertain.

To improve our understanding of radiation-induced mutation rates, both unirradiated and irradiated systems should be studied to obtain more information on the nature of the various forms of genetic damage, their relative frequencies, their modification by repair mechanisms, their heritability and transmission, and their biological (structural and physiological) consequences. Genetic studies at the molecular and cellular level can examine these issues, and the information obtained from them would help investigators reach rational decisions regarding extrapolation to low doses from higher dose data when the effects at low doses are impractical to measure.

Information obtained from studies at the molecular and cellular level will have to be incorporated with that derived from experiments in whole animals and observations in humans, especially since meiosis and gametogenesis in the reproductive process play an important role in the transmission of genetic information.

The nonlinear response to radiation exhibited by gross chromosomal damage is fairly well understood. Some deletions, inversions, and translocations are known to be produced in proportion to the square of the dose for sparsely ionizing radiation, presumably because two chromosome breaks close to one another must be induced before chromosomal rearrangements can take place. The exact relation of this process to the initiation of cancer is not known.

5. *Is the primary molecular lesion produced by ionizing radiation the same whether its expression is cell lethality, mutagenesis, or tumor transformation? If the same primary lesion is responsible for these effects, can any one of these expressions be used to evaluate the effect of radiation on whole organisms, including humans?*

It is tempting to assume that the interaction of ionizing radiation and the cell produces lesions in DNA that cause different observable effects and that the lesions are similar in nature, differing only by their distribution among potential target sites and by number, depending upon dose. If this were true, any convenient effect could be used as a basis for dose-response estimates and their extrapolation to whole organisms, including humans. For such a system, there must be a good theoretical basis for constructing lesion-response curves, a basis that is now lacking because the molecular natures of the lesions are not known.

In addition to the molecular lesions produced by radiation, there must be a reasonably thorough understanding of the cellular factors that serve to modulate the expression of the lesions. Principal modifiers are the relative rates of replication and rates of repair. Cells that have a long time for repair before mitosis or cells that can repair damage very quickly should be less affected than those whose replication rate is rapid or whose repair processes are slow. Thus, it is not sufficient to know the radiation dosimetry at the molecular level alone. Dosimetry data must be correlated with and be considered as a function of cell cycle time.

With respect to the identification of the molecular lesion responsible for the expression of lethality, mutation, or tumor transformation, the current level of knowledge offers some suggestions, but they are often conflicting.

There are a number of direct reasons for associating the *initiating* events in *carcinogenesis* with DNA damage:

- There is a high degree of correlation among compounds that are *mutagenic* when activated appropriately and those that are *carcinogenic*. However, a chemical that reacts with DNA will also readily react with RNA and protein (Ames, 1979), making the target-effect relationship more difficult to assess.
- There is an excellent quantitative correlation between *mutagenicity* and *transformation* in mammalian cell cultures caused by metabolites of polycyclic aromatic hydrocarbons and activated metabolites of nitrosamines (Huberman, 1978; Jones and Huberman, 1980).
- Cells treated *in vitro* with BrdUrd (bromodeoxyuridine) and long-wavelength ultraviolet light, a process known to damage DNA specifically, exhibit neoplastic transformations that correlate with the extent of DNA damage (Barrett *et al.*, 1978).
- UV-irradiation of thyroid cells of the Amazon mollyfish *Poecilia formosa* induces thyroid tumors when the cells are injected into isogenic recipients. However, if the UV-irradiated cells are exposed to a treatment known to monomerize dimers in cellular DNA (photoreactivation), the number of tumors decreases to one-tenth the original (Hart *et al.*, 1977).

• In a number of human disorders, the affected individuals are cancer prone and their cells are more sensitive than normal to exogenous mutagens (Arlett and Lehmann, 1978; Friedberg *et al.*, 1979; Setlow, 1978). Three of these disorders, xeroderma pigmentosum (XP), ataxia telangiectasia (AT), and Fanconi's anemia, are associated with defects in DNA repair systems.

In these cases, however, the association of defective DNA repair and increased cell *cytotoxicity* is weak and is not the same for all individuals in a single disorder group. Moreover, the correlation between the cytotoxic activity of UV and excision repair deficiencies is not a good one (Andrews *et al.*, 1978). This suggests that either there are other repair systems of significance or that cells die for reasons other than the existence of damage to their DNA.

Even in human AT cases, the distribution of cancer types is different than that observed in the population exposed to atomic bomb radiation, further suggesting that the enhanced cancer risk in the AT population might not be attributed to ionizing radiation (Harnden, 1980). Moreover, AT cells are hypomutable to x-irradiation (Arlett and Lehmann, 1978).

• Increases in the *survival* of UV-irradiated cells held in a confluent state before replating correlate with decreases in *mutation* rates and with the rate of excision repair in both proficient and repair-deficient cells (Maher *et al.*, 1979). *Transformation* in human cells also decreases as a result of this procedure (Kakunaga *et al.*, 1980). On the other hand, methods to enhance transformation in Syrian hamster embryo cells do not affect known DNA repair processes (Doniger and Di-Paolo, 1980).

The effects of chemicals and ultraviolet radiation on cells support the concept that DNA is the target for transformation (Setlow, in press). However, the data described above contain contradictory material, and few of them directly involve ionizing radiation. Any inference that mutation, transformation, and cytotoxicity associated with ionizing radiation are quantitatively interchangeable would require evidence of molecular target lesions, their distribution, and their repair in whole cell systems.

If any commonality of mechanisms exists, major stumbling blocks to its demonstration must be overcome. For example, some mutations are caused by base changes in DNA, but others are the result of cytogenetic changes in the chromosomes. Even for the cytogenetic end points, chromosome aberrations, and sister chromatid exchanges (SCEs), there is no simple relationship. The changes in SCE are much greater than the aberrational changes after chemical treatments, but less so after irradiation. Moreover, contrary to what is found for aberrations, there is no effect on SCE if cells are irradiated during the G_2* phase of the cell cycle (Perry and Evans, 1975). These observations imply that chromosomal material responds differently to agents of DNA damage, depending upon the types of lesions induced, the stage of the cell cycle, and possibly even the tertiary organization of the DNA. Therefore, there are at least three modifiers of any concept of a coordinate relationship of cytotoxicity, mutagenicity, and transformation, namely, the spectrum of induced lesions, repair process activity, and possibly DNA organization during the cell cycle.

There is evidence that mortality is correlated with chromosomal aberrations, but a difficulty arises when one attempts to correlate cell killing with molecular events. At a mean lethal dose there may be approximately one aberration per cell, but the initial number of radiation products per haploid genome may be close to 1,000 single-strand breaks, approximately the same number of base damages, 15 double-strand breaks, and at least 100 DNA protein crosslinks (Fornace and Little, 1977). The multiplicity of lesions leads one to ask if a specific unrepaired damage (e.g., a double-strand break) is responsible for the chromosomal abnormality or if the responsible lesion is a random one. Since high-LET radiation has a high RBE for the induction of chromosomal aberrations and induces more double-strand breaks than does low-LET radiation, double-strand breaks have been deemed the lesion most likely to result in the formation of chromosomal aberrations (Evans, 1977; Wolff, 1978). This conclusion is strengthened by the fact that agents such as ultraviolet light and S-dependent chemicals that do not induce double-strand breaks do not produce aberrations other than those formed as errors of replication during the S-phase of the cell cycle, i.e., they do not produce chromosomal aberrations in G_1 or chromatid aberrations in G_2. Additionally, correlations have been made between double-strand breaks themselves and lethality (Ritter *et al.*, 1977). Moreover, chromosomal aberrations that either interfere with cell division or leave the cells genetically imbalanced lead to cell death and are quickly eliminated from the population (Wolff, 1972a).

This interpretation that double-strand breaks and their concomitant

*G_2 is the gap in the interphase of the cell cycle that occurs between the end of DNA synthesis in the "S-phase" and the beginning of mitosis (M).

chromosomal aberrations are the lesions that lead to cell death is consistent with that of Mattern and Welch (1979), who analyzed the excision of one type of base damage—thymine damage induced in DNA by high-LET radiation. They concluded that thymine damage does not contribute to lethality from high-LET exposures. Furthermore, radiations such as ultraviolet light, which do not induce strand breaks, do not induce aberrations in G_1 or G_2 as does ionizing radiation, but only produce lesions that lead to chromatid aberration formation in S-phase by misreplication.

Since chromosomal abnormalities and lethality seem to be associated only with unrepaired or misrepaired DNA strand breaks, it is improbable that there is a close correlation between the lesions involved in killing and those involved in the formation of mutations caused by base changes.

Further evidence that the lesions leading to base change, mutations, and lethality are different comes from the analysis of particular mutations in hamster or human cells. The induction of $HGPRT^-$ (hypoxanthine-guanine-phosphoribosyl transferase) mutations, which could be the result of the loss of the locus, increases linearly with dose in hamsters, but in humans it increases as a higher power of the dose. Both give exponential survival curves (Thacker and Cox, 1975). Moreover, with high-LET radiation (fast neutrons), the RBE for mutations in both humans and hamsters rises to a value of approximately 8, whereas the value for survival is about 4. The shapes of the curves and the high RBEs indicate that these events could be related to chromosomal aberrations such as deletions. However, some specific mutations, such as those for resistance to ouabain, cannot be induced by ionizing radiation, whereas they can be produced by ultraviolet. The lethality of deletions at the ouabain locus provides further evidence that x-ray-induced mutations are largely chromosomal deletions, even in the mouse (Abrahamson and Wolff, 1976; Russell, 1980).

Although certain types of mutational effects can be related to cell killing, there is no clear way to correlate mutagenesis with transformation. The attempted correlation is difficult because the expression of mutation induction in mammalian cells involves relatively long expression periods after irradiation. Transformation involves much longer times and subculturing (Borek, 1980), and even *in vitro* may involve the action of promoters and inhibitors (Borek *et al.*, 1979; Kennedy *et al.*, 1978). Such steps are not involved in the production of mutagenic or even cytotoxic end points.

Some experiments indicate a quantitative relationship between cell survival and cell transformation. Terzaghi and Little (1976) irradiated confluent cultures of mouse embryo cells. The cultures were diluted and replated to measure survival and the number of transformants per survivor. When the cells were diluted approximately 6 h after irradiation, both survival and transformation increased. After that, transformation dropped off markedly, but survival did not. Thus, there is a seeming divergence of response between survival and transformation. If the same lesions were involved in both processes, one would predict, naively, that if survival increased (equivalent to response to a smaller dose) transformation would decrease, but this does not take place. The authors hypothesize that errors in repair of damage are implicated in the neoplastic transformation. There is no way of testing this hypothesis until the nature of the lesions is understood. We do know that any correlations observed between survival and transformation were not strictly quantitative.

Miller *et al.* (1979) provide additional clues indicating that *survival* and *transformation* have little in common. Split doses between about 25 and 200 rad administered 5 h apart produced significantly more transformants than a single dose equal to the sum of the two split ones. At higher doses, the split regimen provided fewer transformants. This crossover in transformation effect does not appear in the survival curves for which the split dose always results in higher survival. Either the lesions are different or the biology still eludes us.

As noted above, it is fashionable to attempt to explain many of the observed phenomena in terms of repair. For example, *a priori* one might have thought the higher yield of radiation-induced chromosome breaks in Down's syndrome to be the result of a slower restitution of the broken ends, resulting in a longer time for the chromosomes to find the wrong partners. However, Countryman *et al.* (1977) assessed this hypothesis with split dose experiments and showed that Down's syndrome cells work faster at restituting breaks. The authors suggest that this fast system may be error-prone, a proposal that needs to be tested.

6. *What factors determine the extrapolation of data on molecular effects to the effects on cells?*

It is clear from the previous discussions that before data can be extrapolated from the molecular effects of radiation to effects on the whole cell, we must at least know which molecular effects are important for each of the separate end points and the expression times for mutation, transformation, and lethality in terms of dose characteristics and cellular physiology (as well as the role of promotors or inhibitors related to each end point). Two extreme classes of change can be considered. One class is dose-dependent, i.e., increased doses would affect either an increased number of cells or produce greater effects on all of the cells. In either case, the change would be expected to be completed when the dose is sufficient to affect all cells or to affect each cell to its maximum capacity—a saturation level. This type of change is found in the induced release of virus particles from cells irradiated with ionizing radiation (Shinagawa *et al.*, 1977) and in the induction of enzyme systems (see Hanawalt *et al.*, 1979, for UV induction of repair enzymes in bacteria and mammalian cells). In both of these examples, many or all of the exposed cells experience a similar change in response to radiation.

A second class of change includes those alterations that differ from one cell to another. A wide variety of possible discrete changes, such as mutations, is possible. Any single cell may or may not have experienced one or another of the changes, and the probability of finding a cell with a specific change would be expected to be dose-dependent. Such changes resulting from exposures to UV-irradiation and chemicals include mutations that are the result of DNA damage or flawed repair of the lesions. The heritability of such changes permit their identification as mutants. Some such changes may be sufficiently drastic to compromise the viability of the cell. Lethal effects of radiation exposure may include a large array of these drastic changes, each different and each lethal.

Under these conditions, efforts at extrapolation require knowledge of the number and types of changes and their consequences for biological reactions such as DNA replication and RNA transcription.

Much more insight into these issues is necessary because changes in DNA and in other cellular structures are not necessarily stable, but may change with time. Consideration must be given to the rates of spontaneous decay of lesions, rates of repair, and rates of cell replication and of DNA transcription.

In mammalian systems, repair of excisions and strand breaks occurs in nondividing cells. One might suspect that there would be great difficulty in extrapolating from observations made following a low chronic dose of ionizing radiation without having detailed information on the distribution of cells in the cell cycle and of cycling cells in the exposed tissue.

The extrapolation to whole cells from the effects of radiation on cellular components is complicated by the existence of many possible chemicals that may be promoters or antipromoters in neoplastic transformation. The identity of these agents must be known, and their modes of action understood. Neither of these requirements is met by our current level of knowledge.

Although to this point we have assumed that one of the effects of radiation, carcinogenesis, may result from a mutagenic event, an additional feature should be considered. Nonmutational alterations due to ionizing radiation are common. Such alterations, when transmitted to progeny cells can be referred to as epigenetic changes. They need not themselves be neoplastic changes, but could result in an enhanced probability of neoplastic change later in the history of the progeny cells (e.g., in mouse cells exposed to ionizing radiation; Kennedy and Weichselbaum, 1981; Kennedy *et al.*, 1980). The nature of this change is not understood, but it could be a modifier of enzyme systems or an alteration in state of a latent virus carried by the cell.

An additional level of complexity in understanding the system that further confounds our capacity to extrapolate from molecular lesions to the cell lies in the amplifying effects of certain chemical agents or hormones on x-ray-induced tumors (Kennedy *et al.*, 1980). The extent to which these substances affect the probability of occurrence of neoplastic transformation or affect the growth rate of the neoplastic cell once it has appeared is not yet clear.

All of these variables, as well as others not as yet identified, should be understood before extrapolation from molecule to cell can be meaningfully undertaken.

7. *What factors determine the extrapolation of data on cellular effects to the effects on whole organisms?*

Whereas the development of an understanding of the molecular effects of ionizing radiation seems to be an attainable goal through aggressive study, and modifying factors in the cellular environment may gradually be identified, the relationships of the total cell to the total organism impose a level of complexity that is a far greater task to unravel.

One approach to dealing with the extrapolation from cell to organism is to consider the available information and the factors associated with each major cellular-molecular aberration due to ionizing radiation (National Academy of Sciences, 1980).

Cell lethality has been discussed above. The reasons for the differing sensitivities of cells of different organs (e.g., different rates of cell proliferation) are only partly understood. Hence, the impact of the loss of cellular viability on the whole organism will vary depending upon the organ or tissue most seriously affected and the metabolic functions that are compromised (Bergonié and Tribondeau, 1959; Fabrikant, 1972; Wald, 1975). A determining factor would be the recovery potential of the damaged organs, which would be partially reflected in the cell replacement capacity of each organ. These considerations probably play a prominent role in the effects of ionizing radiation on the developing embryo (Murphree and Pace, 1960; Nøkkentved, 1968; Russell *et al.*, 1960).

Mutagenesis is ultimately responsible for the appearance of cells that are different from the cells of the tissue of origin. The extent of these differences may vary widely, ranging from subtle or inconsequential changes that do not perturb the organism to extreme changes resulting in lethality and the loss of the capacity of the organism to multiply. Between these extremes probably lies a wide variety of cellular changes that affect the survival and reproduction of organisms. Most of the lethal effects of radiation exposure can be accounted for by genetic damage that results in loss or rearrangement of chromosomes (Brewen and Preston, 1975; Cacheiro *et al.*, 1974; Cox and Lyon, 1975; Wolff, 1972b).

Mutational events in germline cells may lead to destruction of the cells, abnormalities that prevent their utility in reproductive processes, or abnormalities that appear in subsequent generations. Evaluation of these mutational effects must be based upon knowledge of the targets of radiation injury, their sensitivity to damage, the range of genetic alterations and their consequences, and the relationship of all these factors to the unique steps of germinal cell development, meiosis, and gametogenesis, which may alter the sensitivity of those cells to change. The consequences of mutations occurring in somatic cells are poorly understood. It has been suggested that the accumulation of cells with such mutations could account for aging. Perhaps more to the point, in malignant transformation, carcinogenesis seems likely as a consequence of such mutations. Although neither the molecular events nor the cellular changes are understood, the *carcinogenic effects* of exposure to ionizing radiation have been clearly demonstrated (National Academy of Sciences, 1980). Cell growth and cell loss within an organ are tightly regulated as evidenced by the fact that despite continuous replacement of cells within an organ, the organ does not change much in size during adult life. Neoplasms are presumed to arise as single clones derived from a cell that has experienced either a loss or an alteration of its normal growth constraints (Fialkow, 1974).

Many models have been suggested to describe the sequence of

events that results in the appearance of a malignant tumor (Cairns, 1978; Peto, 1977). Some of the models include a step involving a DNA change, i.e., a mutational step. Such a mutational step may indeed represent the connection between radiation exposure and carcinogenesis. Several features of mammalian cells grown in culture make them an attractive model system for the investigation of carcinogenesis. An alternative model is represented by the radiation-induced activation of a latent oncogenic virus that leads to malignancy as demonstrated in mouse leukemias and lymphomas and osteogenic sarcoma.

When isolated mammalian cells are grown in culture, they adhere to the culture vessel and multiply by spreading over the surface of the vessel. When the cell number increases to the point that the surface of the vessel is covered by a confluent monolayer of cells, cell multiplication ceases. The cells exhibit contact inhibition. For some cell lines, exposure of cells to ionizing radiation results in such inhibition after the cells have grown into a confluence of colonies that are many cells thick. The cells of these colonies are *transformed* in that they no longer exhibit contact inhibition (Borek, 1980; Terzaghi and Little, 1976). In many cases, cells from such colonies give rise to malignant tumors when injected into test animals (Terzhagi and Little, 1976). The yield of transformed clones from a population of cells exposed to ionizing radiation is dose-dependent over a limited range (Terzhagi and Little, 1976).

Promotion can be defined as a treatment, either chemical or physical, that itself would not be carcinogenic, but when applied to animals that had received prior exposure to a carcinogen will amplify the impact of that exposure (Boutwell, 1974). In mouse skin-painting experiments, large numbers of tumors will appear if single exposures to a carcinogen such as dimethylbenzanthracene, at a level sufficient to induce only a few tumors, are followed by repeated treatments with a promoter such as phorbol ester. Those tumors will appear after a shorter latent period than tumors induced by exposure to the carcinogen alone (Boutwell, 1974). It is presumed that the exposure to a carcinogen has resulted in mutational alterations and that in some way promotion enhances "uncontrolled growth leading to cancer" on the part of a mutationally altered cell. Although there are many speculations concerning the molecular and cellular events that are relevant to promotion, a clear picture has not yet emerged. Since promotion may be responsible for a very large increase in tumor incidence following exposure to a carcinogen, studies to further the understanding of this process must receive high priority.

Cells in culture provide an excellent material for probing the mechanisms of transformation and perhaps of promotion. Contact inhibition appears to bear some resemblance to organ-specific size regulation in the mature animal. Exposure of cells in culture to ionizing radiation results in the appearance of transformed clones whose cells no longer show contact inhibition and are capable of forming malignant tumors in certain test animals. This system bears some resemblance to the x-ray induction of tumors in the skin of rats (Albert *et al.*, 1967). In addition, studies of the cell culture system is a means of approaching an understanding of the role of promotion in tumorigenesis. The radiation-induced yield of transformed colonies may be enhanced when cells that have been exposed to ionizing radiation are treated with substances that display promotional activity in animals.

A number of cellular phenomena require elucidation if the carcinogenesis induced by exposure to ionizing radiation is to be understood. Some of these can be reasonably well characterized and can probably best be investigated using the approaches of molecular and cell biology. Although it is attractive to assume that the primary effect of radiation on the induction of transformed cells in culture is a mutational change, this remains to be demonstrated. Reports of "cell density effects" claim an enhanced yield of transformants when reduced densities of cells that have been exposed to a given radiation dose are assayed (Terzhagi and Little, 1976). Furthermore, it has been reported that the cells destined to exhibit the transformed property have not expressed their new property even 12 generations after exposure to radiation (Terzhagi and Little, 1976). These observations suggest that a mutational change is not sufficient to account for the phenomenon. Investigations of bacteria have indicated that exposure to UV radiation can result in the induction of repair enzymes in the exposed population (Kenyon and Walker, 1980).

Both the cell density effects and the delay in expression of the transformant property could be accounted for if one were to assume that a radiation-induced regulatory change persisted in the clonal progeny of the irradiated cell and that this change endowed the progeny with an enhanced probability of the appearance of a transformed cell many generations later. Although speculative, the current limited understanding of the overall process allows for the consideration of such a possibility (Kennedy *et al.*, 1980).

In examining the role of promotion, the notion of a growth rate enhancement of altered cells seems attractive but remains to be demonstrated (Farber and Cameron, 1980). The understanding of the effects on cells in culture induced by exposure to ionizing radiation is central to understanding the effects on whole organisms. In addition to understanding the effects on DNA and protein, it is important that we understand the variety of cell responses and their patterns of dose dependence. These include mutagenicity, changes in regulation of synthesis of certain proteins, the capacity of cells to repair damages, the fidelity with which the damages are repaired, and the impact of radiation-induced changes on the growth patterns of cells that have survived exposure to radiation.

8. *Why is there a latent period between radiation exposure and the observed effect?*

A latent period between exposure and the expression of detectable effect is characteristic of a variety of effects, but has different meanings, depending upon the result, the method of detection, and the tissue or organ of interest. It is defined as the "period of seeming inactivity between the time of exposure of tissue to an injurious agent and response" (National Academy of Sciences, 1980).

Degenerative changes such as cataractogenesis involve damage to the specialized epithelial cells that cover the lens. Latency in this case varies from months to years and is dose-dependent, being shorter at higher doses (Merriam and Focht, 1957; National Council on Radiation Protection and Measurements, 1971). More information on the general pathogenesis of cataracts is needed, especially on the progression from minor opacification to clinically significant cataract formation. The host factors that might influence the development of cataracts must also be better understood. This is yet another example of the utility of studying a biological problem with and without respect to ionizing radiation, since the initiating event may not be the controlling factor in the progress of the abnormality.

Latency of the effects of radiation on male *fertility* is related to cell kinetics and the differentiation sequence for spermatogenesis, to sperm storage, and to reproductive activity. Latency of the female germinal cell abnormalities—infertility and sterility—pertains only to the time of observation since killing or irreparable injury to the full complement of oocytes present early in life is probably almost immediate. Thus, the existence of the change is recognized only when subsequent reproductive activity is attempted. Similarly, latency of *genetically determined birth defects* spans the time between injury of a gamete and the development of progeny to a stage at which the abnormality may be observed.

Developmental defects depend on the time during prenatal life that exposure to an injurious agent occurs. They are related to the stage of prenatal development, the degree of differentiation of tissue cells, the cell cycle periods, and the number of cycling cells in susceptible stages in the target zone. The latency period in this case is the time necessary for the development of structure or function to the point that abnormalities can be recognized. These abnormalities provide dramatic examples of the need to understand the relationship of molecular lesions induced by ionizing radiation in somatic cells to the state of the target cells, e.g., cells at various points in the cell cycle. Clarification of the stage at which the target cell is associated with either the presence or absence of an abnormality of later tissue development would be the most useful biological information to apply to developmental abnormalities and to cancer induction.

The latency associated with the impact of ionizing radiation on the development of *tumors* has a more complex set of characteristics. Since the period between exposure and tumor development may vary from 2 to 15 or more years, depending upon the tissue exposed and the specific tumor developed, many sets of functions must be considered in addition to those listed for the other effects of radiation. First, the fundamental, as yet unanswered, question must be addressed: How is the molecular change following irradiation of whole tissues and whole organisms finally expressed as an autonomously growing cell mass? In addition, the following questions must be answered:

- If carcinogenesis is a multistage process, as has been hypothesized, what events must occur, and at what intervals, to yield a tumor?
- Which host factors affect the success or failure of a cell prepared to become a tumor? Do local cell-to-cell communication systems afford an essential control mechanism? How do hormonal, immunological, nutritional, neurogenic, and other systems influence the development of a tumor?
- Is ionizing radiation an initiator? Could it act as a promoter? Or both? Does it act only through the physicochemical changes it induces in the cell by virtue of its energy impact, or does it act, as noted previously, to permit the expression of a precedent abnormality, e.g., latent virus?
- Are there promoters and antipromoters in the cellular environment? How do they act?
- What molecular, cellular, and control conditions influence the

differences in tissue response to ionizing radiation in terms of latent periods?

• Can molecular repair capacity be equated with control of initiation or promotion?

• What is the significance of age and sex of the organism in cancer induction and latent periods? Does the relative risk projection (National Academy of Sciences, 1980) have an identifiable biological basis?

It is obvious from these selected questions that the basic biology of carcinogenesis requires study in order to answer questions about ionizing radiation. Conversely, ionizing radiation can be used as a probe of the biology of carcinogenesis.

9. *What factors determine the distribution of internal radiation emitters in tissue, and how are they related to their biological effectiveness? How well are these factors understood, and what are the important unanswered questions concerning them?*

Each radionuclide has a unique decay rate and set of radiations, and each exhibits the properties of the chemical of which it is an isotope. The chemistry of a radioelement determines how it will move in the environment, the degree to which it will be taken into the body, the tissues in which it will be deposited, and, therefore, the tissues to which its radiation energy will be delivered. Most radionuclides irradiate only part of the body, and there is great variety in their dose-rate patterns and the amounts of energy delivered per unit of path length of the emitted particles. That variety makes radionuclides versatile scientific tools, but it makes measurement of the radiation dose to a tissue a complex matter. For example, work begun 30 yr ago is still in progress to reconstruct the radiation doses received by the radium dial painters, uranium miners, and Thorotrast-treated patients—all important human populations exposed to internally deposited radionuclides.

Studies have been conducted to investigate the physiological behavior of radionuclides—element by element—and to study the biological effects of several important members of this group. Radionuclides may occur in many chemical forms and enter the body by inhalation, ingestion, or through broken skin. Thus, it has been necessary to study the transport of a variety of radioactive compounds by all routes of entry.

The doses to tissues from radionuclides usually cannot be measured directly but must be calculated for each nuclide in each tissue. The effective dose in a tissue from a deposited radionuclide is proportional to the intake and the energy/disintegration/unit weight of tissue, modified by a number of factors such as the fraction of the intake absorbed, the fraction of the body content initially deposited in the tissue, the fraction of the initial deposit retained at the time of interest, and the fraction of the emitted energy absorbed in the tissue. Most investigators studying radionuclide absorption have used compounds likely to give the highest or lowest values in an attempt to reduce the effort needed to obtain useful results. The deposition and retention functions are not known for all radioelements in human tissues, and for a given radionuclide they are not always the same in all the mammals studied.

Much remains to be learned about the comparative biochemistry and the rates of cell renewal of different mammals, processes believed to underlie, at least in part, species differences in the deposition and retention of radionuclides. The mass of cells that absorb the energy emitted by a radionuclide and, perhaps, later develop a cancer is not always known because the critical cells have not been identified. In several important cases, only part of the energy emitted by a radionuclide deposited in that tissue is absorbed in its cells. A correction for the "wasted" energy must be based on the sizes and shapes of the collections of critical cells. Such measurements in animals and humans are required to extrapolate data on radiation dose and biological effects derived from animals to the assessment of radiation risks in humans when there are no such data for humans.

The radiation-protection system newly devised by the International Commission on Measurements and Units and the International Commission on Radiological Protection (1977) requires data for the deposition and retention of many radionuclides in tissues that heretofore were not investigated. Nearly all of the biological data needed to solve the radiation dose equations used for protection purposes were developed for the adult male—the typical radiation worker (International Commission on Radiological Protection, 1975). To extend dose calculations to protect a general population, new metabolic and biological data need to be acquired for women and children.

Nearly all of the radiobiological studies of radionuclides in animals have used a single intake, which is characterized by an exponentially declining dose rate in the tissues. Yet, the intake pattern for environmentally dispersed radionuclides is projected to be nearly constant over years or a lifetime—like that of naturally occurring radionuclides. Metabolic studies of populations living in areas of high natural radionuclide content and the measurements of environmental levels and human bone and tissue contents of fallout strontium-90 and plutonium-239 provide useful information for dose calculations, but are of limited value in deriving dose-effect relationships. Therefore, there is a need for animal studies in which radionuclides are administered repeatedly at low dosages over an extended time to approximate continuous intake.

SUMMARY

• Further studies with increased emphasis on understanding the mechanisms of radiation carcinogenesis should be encouraged. This is particularly important with respect to the making of risk estimates at low doses of low-LET radiation. This research should involve cellular and molecular experiments combined with selected observations on irradiated animals and appropriate results from irradiated human populations. The design and conduct of experiments that test current concepts and models of carcinogenesis in general and radiation carcinogenesis in particular should also be encouraged.

• Radiation poses not only somatic risks, such as the induction of cancer, but also genetic risks, which can affect future generations. These risks have been quantified, to the limits of present knowledge, in the BEIR III report (National Academy of Sciences, 1980). Because of the uncertainties surrounding the nature of the various forms of genetic damage, their modification, their heritability, their transmission, and their biological consequences, the estimates of risk to humans are imprecise. Consequently, studies in radiation genetics addressing the uncertainties must continue, particularly at the molecular and cellular level. However, because reproductive processes such as meiosis and gametogenesis play a large role in the transmission of genetic damage, experiments still need to be carried out on whole organisms. In addition, studies, particularly at very low doses, will also have to be conducted on single cells from animals and plants and on single-celled organisms.

• In addition to the need for a basic understanding of mechanisms to aid in evaluating risk, further observations on the effects on whole animals are required. These observations should be directed toward evaluating effects on sensitive subpopulations, such as developing fetuses, newborn animals, and organisms, and studying the physiological and metabolic processes that determine the dose distribution from both internal and external radiation sources.

REFERENCES

Abrahamson, S., and S. Wolff. 1976. Re-analysis of radiation-induced specific locus mutations in the mouse. Nature 264:715-719.

Abrahamson, S., M. A. Bender, A. D. Conger, and S. Wolff. 1973. Uniformity of radiation-induced mutation rates among different species. Nature 245:460-462.

Albert, R. E., F. J. Burns, and R. D. Heimbach. 1967. The association between chronic radiation damage of the hair follicles and tumor formation in the rat. Radiat. Res. 30:590-599.

Ames, B. N. 1979. Identifying environmental chemicals causing mutations and cancer. Science 204:587-593.

Andrews, A. D., S. F. Barrett, and J. H. Robbins. 1978. Xeroderma pigmentosum neurological abnormalities correlate with colony forming ability after ultraviolet radiation. Proc. Natl. Acad. Sci. USA 75:1984-1988.

Arlett, C. F., and A. R. Lehmann. 1978. Human disorders showing increased sensitivity to the induction of genetic damage. Annu. Rev. Genet. 12:95-115.

Barrett, J. C., T. Tsutsui, and P. O. P. Ts'o. 1978. Neoplastic transformation induced by a direct perturbation of DNA. Nature 274:229-232.

Bergonié, J., and L. Tribondeau. 1959. Interpretation of some results of radiotherapy and an attempt at determining a logical technique of treatment. (Translation of paper originally published in 1906: Compt. Rend. 143:983-985.) Radiat. Res. 11:587-588.

Bond, V. P., C. J. Shellabarger, E. P. Cronkite, and T. M. Fliedner. 1960a. Studies on radiation-induced mammary gland neoplasia in the rat. 5. Induction by localized irradiation. Radiat. Res. 13:318-328.

Bond, V. P., E. P. Cronkite, S. W. Lippincott, and C. J. Shellabarger. 1960b. Studies on radiation-induced mammary gland neoplasia in the rat. 3. Relation of the neoplastic response to dose of total-body radiation. Radiat. Res. 12:276-285.

Borek, C. 1980. X-ray-induced *in vitro* neoplastic transformation of human diploid cells. Nature 283:776-778.

Borek, C., R. Miller, C. Pain, and W. Troll. 1979. Conditions for inhibiting and enhancing effects of the protease inhibitor antipain on x-ray-induced neoplastic transformation in hamster and mouse cells. Proc. Natl. Acad. Sci. USA 76:1800-1803.

Boutwell, R. K. 1974. The function and mechanism of promoters of carcinogenesis. CRC Crit. Rev. Toxicol. 2:419-443.

Brewen, J. G., and R. J. Preston. 1975. Analysis of x-ray induced chromosomal translocations in human and marmoset spermatogonial stem cells. Nature 253:468-470.

Cacheiro, N. L. A., L. B. Russell, and M. S. Swartout. 1974. Translocations, the predominant cause of total sterility in sons of mice treated with mutagens. Genetics 76:73-91.

Cairns, J. 1978. The epidemiology of cancer. Pp. 35-61 in Cancer: Science and Society. W. H. Freeman and Co., San Francisco.

Cairns, J. 1981. The origin of human cancers. Nature 389:353-357.

Countryman, P. I., J. A. Heddle, and E. Crawford. 1977. The repair of x-ray-induced chromosomal damage in trisomy 21 and normal diploid lymphocytes. Cancer Res. 37:52-58.

Cox, B. D., and M. F. Lyon. 1975. X-ray induced dominant lethal mutations in mature and immature oocytes of guinea pigs and golden hamsters. Mutat. Res. 28:421-436.

Doniger, J., and J. A. DiPaolo. 1980. Excision and postreplication DNA repair capacities, enhanced transformation, and survival of Syrian hamster embryo cells irradiated by ultraviolet light. Cancer Res. 40:582-587.

Evans, H. J. 1977. Molecular mechanisms in the induction of chromosome aberrations. Progr. Genet. Toxicol. 2:57-74.

Fabrikant, J. I. 1972. Radiation effects on rapidly and slowly proliferating cell renewal systems. Pp. 57-78 in J. M. Vaeth, ed. Frontiers of Radiation Therapy and Oncology. Vol. 6. S. Karger, Basel.

Farber, E., and R. Cameron. 1980. The sequential analysis of cancer development. Cancer Res. 31:125-226.

Fialkow, P. J. 1974. The origin and development of human tumors studied with cell markers. N. Engl. J. Med. 291:26-35.

Fornace, A. J., Jr., and J. B. Little. 1977. DNA crosslinking induced by x-rays and chemical agents. Biochim. Biophys. Acta 477:343-355.

Friedberg, E. C., U. K. Ehmann, and J. I. Williams. 1979. Human diseases associated with defective DNA repair. Adv. Radiat. Biol. 8:85-174.

Hanawalt, P. C., P. K. Cooper, A. K. Ganesan, and C. A. Smith. 1979. DNA repair in bacteria and mammalian cells. Annu. Rev. Biochem. 48:783-836.

Harnden, D. G. 1980. Mechanisms of genetic susceptibility to cancer. Pp. 235-244 in B. Pullman, P. O. P. Ts'o, and H. Gelboin, eds. Carcinogenesis: Fundamental Mechanisms and Environmental Effects. D. Reidel, Dordrecht.

Hart, R. W., R. B. Setlow, and A. D. Woodhead. 1977. Evidence that pyrimidine dimers in DNA can give rise to tumors. Proc. Natl. Acad. Sci. USA 74:5574-5578.

Hollaender, A., ed. 1954. Radiation Biology. Vol. I: High Energy Radiation. Prepared under the auspices of the National Research Council, National Academy of Sciences, Washington, D.C. McGraw-Hill Book Co., Inc., New York. 1,265 pp.

Huberman, E. 1978. Mutagenesis and cell transformation of mammalian cells in culture by chemical carcinogens. J. Environ. Pathol. Toxicol. 2:29-42.

International Commission on Radiological Protection. 1975. Recommendations of the International Commission on Radiological Protection. ICRP Publication 23. Pergamon Press, Oxford. 480 pp.

International Commission on Radiological Protection. 1977. Recommendations of the International Commission on Radiological Protection. ICRP Publication 26. Pergamon Press, Oxford. 53 pp.

Jacobson, L. O., E. K. Marks, E. O. Gaston, M. Robson, and R. E. Zirkle. 1949. The role of the spleen in radiation injury. Proc. Soc. Exp. Biol. Med. 70:740-745.

Jones, C. A., and E. Huberman. 1980. A sensitive hepatocyte-mediated assay for the metabolism of nitrosamines to mutagens for mammalian cells. Cancer Res. 40:406-411.

Kakunaga, T., K.-Y. Lo, J. Leavitt, and M. Ikenaga. 1980. Relationship between transformation and mutation in mammalian cells. Pp. 527-541 in B. Pullman, P. O. P. Ts'o, and H. Gelboin, eds. Carcinogenesis: Fundamental Mechanisms and Environmental Effects. D. Reidel, Dordrecht.

Kellerer, A. M., and H. H. Rossi. 1972. The theory of dual radiation action. Curr. Topics Radiat. Res. 8:85-158.

Kennedy, A. R., and R. Weichselbaum. 1981. Effects of 17β-estradiol on radiation transformation *in vitro*; Inhibition of effects by protease inhibitors. Carcinogenesis. 2:1:67-69.

Kennedy, A. R., S. Mondal, C. Heidelberger, and J. B. Little. 1978. Enhancement of x-ray transformation by 12-0-tetradecanoyl-phorbol-13-acetate in a cloned line of C3H mouse embryo cells. Cancer Res. 38:439-443.

Kennedy A. R., M. Fox, G. Murphy, and J. B. Little. 1980. Relationship between x-ray exposure and malignant transformation in C3H 10T½ cells. Proc. Natl. Acad. Sci. USA 77:7262-7266.

Kenyon, C. J., and G. C. Walker. 1980. DNA-damaging agents stimulate gene expression at specific loci in *Escherichia coli*. Proc. Natl. Acad. Sci. USA 77:2819-2823.

Lea, D. E. 1946. Actions of Radiations on Living Cells. Cambridge University Press, Cambridge, England. 402 pp.

Lorenz, E., C. Congdon, and D. Uphoff. 1952. Modification of acute irradiation injury in mice and guinea-pigs by bone marrow injections. Radiology 58:863-877.

Maher, V. M., D. J. Dorney, A. L. Mendrala, B. Konze-Thomas, and J. J. McCormick. 1979. DNA excision-repair processes in human cells can eliminate the cytotoxic and mutagenic consequences of ultraviolet irradiation. Mutat. Res. 62:311-323.

Mattern, M. R., and G. P. Welch. 1979. Production and excision of thymine damage in the DNA of mammalian cells exposed to high-LET radiations. Radiat. Res. 80:474-483.

Mays, C. W. 1976. Estimated risk from Pu 239 to human bone, liver and lung. Pp. 373-384 in Biological and Environmental Effects of Low-Level Radiation. Vol. 2. Proceedings of a Symposium, Chicago, 3-7 November, 1975, organized by the IAE and WHO. International Atomic Energy Agency, Vienna.

Mays, C. W., and M. P. Finkel. 1980. RBE of α-particles vs. β-particles in bone sarcoma induction. Pp. 401-405 in Proceedings of the Fifth International Congress of IRPA, Jerusalem, Israel, Vol. II.

Mays, C. W. and R. D. Lloyd. 1972. Bone sarcoma risk from ^{90}Sr. Pp. 352-375 in M. Goldman and L. K. Bustad, eds. Biomedical Implications of Radiostrontium Exposure. Proceedings of a symposium held at Davis, Calif., Feb. 22-24, 1971, sponsored by Radiobiology Laboratory, University of California, and U.S. Atomic Energy Commission. AEC Symposium Series 25, CONF-710201. National Technical Information Service, Springfield, Va.

Merriam, G. R., Jr., and E. F. Focht. 1957. A clinical study of radiation cataracts and the relationship to dose. Am. J. Roentgen. 77:759-785.

Miller, R. C., E. J. Hall, and H. H. Rossi. 1979. Oncogenic transformation of mammalian cells *in vitro* with split doses of x-rays. Proc. Natl. Acad. Sci. USA 76:5755-5758.

Muller, H. J. 1927. Artificial transmutation of the gene. Science 66:84-87.

Muller, H. J. 1928. The production of mutations by x-rays. Proc. Natl. Acad. Sci. USA 14:714-726.

Murphree, R. L., and H. B. Pace. 1960. The effects of prenatal radiation on postnatal development in rats. Radiat. Res. 12:495-504.

National Academy of Sciences. 1980. The Effects on Populations of Exposure to Low Levels of Ionizing Radiation. Report of the Committee on the Biological Effects of Ionizing Radiations. National Academy of Sciences, Washington, D.C. 638 pp.

National Council on Radiation Protection and Measurements. 1971. Basic Radiation Protection Criteria. NCRP Report No. 64. National Council on Radiation Protection and Measurements, Washington, D.C. 135 pp.

National Council on Radiation Protection and Measurements. 1980. Influence of Dose and Its Distribution in Time on Dose-Response Relationships for Low-LET Radiations. Recommendations of the NCRP. NCRP Report No. 64. National Council on Radiation Protection and Measurements, Washington, D.C.

Nøkkentved, K. 1968. Effect of Diagnostic Radiation upon the Human Foetus. Munksgaard, Copenhagen. 222 pp.

Perry, P., and H. J. Evans. 1975. Cytological detection of mutagen-carcinogen exposure by sister chromatid exchange. Nature 258:121-125.

Peto, R. 1977. Epidemiology, multistage models, and short-term mutagenicity tests. Pp. 1,403-1,428 in H. H. Hiatt, J. D. Watson, and J. A. Winsten, eds. Origins of Human Cancer, Book C, Human Risk Assessment. Cold Spring Harbor Conference on Cell Proliferation. Vol. 4. Cold Spring Harbor Laboratory, Long Island, N.Y.

Ritter, M. A., J. E. Cleaver, and C. A. Tobias. 1977. High-LET radiations induce a large proportion of non-rejoining DNA breaks. Nature 266:653-655.

Russell, L. B. 1980. Repair and mutation induction in mouse germ cells: A summary and some thoughts. Pp. 421-427 in W. M. Generoso, M. D. Shelby, and F. J. de Serres, eds. DNA Repair and Mutagenesis in Eukaryotes. Plenum Press, New York.

Russell, L. B., S. K. Badgett, and C. L. Saylors. 1960. Comparison of the effects of acute, continuous, and fractionated irradiation during embryonic development. Pp. 343-359 in A. A. Buzzati-Traverso, ed. Immediate and Low-Level Effects of Ionizing Radiations. Proceedings of the symposium held at Venice, June 22-24, 1959, under the joint sponsorship of UNESCO, IAEA, and CNRN and published as a supplement to the International Journal of Radiation Biology. Taylor and Francis Ltd., London.

Samson, L., and J. Cairns. 1977. A new pathway for DNA repair in *Escherichia coli*. Nature 267:281-283.

Samson, L., and J. L. Schwartz. 1980. Evidence for an adaptive DNA repair pathway in CHO and human skin fibroblast cell lines. Nature 287:861-863.

Setlow, R. B. 1978. Repair deficient human disorders and cancer. Nature 271:713-717.

Setlow, R. B. In press. DNA damage and carcinogenesis in chromosome damage and DNA repair. Pp. 575-584 in E. Seeberg and K. Kleppe, eds. Chromosome Damage and Repair. NATO ASI Series A: Life Sciences Section. Plenum Publishing Corp., New York.

Shellabarger, C. J., D. Chmelevsky, and A. M. Kellerer. 1980. Induction of mammary neoplasms in the Sprague-Dawley rat by 430-keV neutrons and x-rays. J. Natl. Cancer Inst. 64:821-833.

Shinagawa, H., K. Mizuuchi, and P. T. Emmerson. 1977. Induction of prophage lambda by γ-rays, Mitomycin C and *tif*; repressor cleavage studied by immunoprecipitation. Molec. Gen. Genet. 155:87-91.

Terzaghi, M., and J. B. Little. 1976. X-radiation-induced transformation in a C3H mouse embryo-derived cell line. Cancer Res. 36:1367-1374.

Thacker, J., and R. Cox. 1975. Mutation induction and inactivation in mammalian cells exposed to ionizing radiation. Nature 258: 429-431.

Upton, A. C., F. F. Wolff, J. Furth, and A. W. Kimball. 1958. A comparison of the induction of myeloid and lymphoid leukemias in x-radiated RF mice. Cancer Res. 18:842-848.

Wald, N. 1975. Radiation injury. Pp. 66-72 in P. B. Beeson and W. McDermott, eds. Textbook of Medicine, 14th ed. W. B. Saunders Co., Philadelphia.

Wolff, S. 1972a. Genetic effects and radiation-induced cell death. Pp. 459-469 in J. M. Vaeth, ed. Frontiers of Radiation Therapy and Oncology. Vol. 6. Radiation Effect and Tolerance, Normal Tissue Basic Concepts in Radiation Pathology. Proceedings of the 6th Annual San Francisco Cancer Symposium, 1970. S. Karger, Basel.

Wolff, S. 1972b. The repair of x-ray-induced chromosome aberrations in stimulated and unstimulated human lymphocytes. Mutat. Res. 15:435-444.

Wolff, S. 1978. Relation between DNA repair, chromosome aberrations, and sister chromatid exchanges. Pp. 751-760 in P. C. Hanawalt, E. C. Friedberg, and C. F. Fox, eds. DNA Repair Mechanisms. Academic Press, New York.

7
Transport Systems and Ecology

The consequences to humans, to other living species, and to ecosystems resulting from the release of man-made radionuclides into the environment, or even from the distribution and uptake of naturally occurring radionuclides, depend on many factors. The impact of these radionuclides is influenced by the following characteristics:

- factors at the source including release rates or probabilities, physical and chemical properties of the radionuclides, such as chemical forms or particle sizes, and associated materials released simultaneously;
- dispersal characteristics, including atmospheric and hydrological processes that determine the distribution and mobility of the radionuclides in the environment;
- biotic processes, which influence the biological availability, routes of uptake, and concentration rates of the dispersed radionuclides; and
- physiological mechanisms, which determine the uptake of the radionuclides, their retention, translocation to critical organs, and the radiosensitivity to the inhaled or ingested radionuclides.

The potential health and other biological effects of environmental radionuclides are explained in Chapters 4, 5, and 6. They are influenced by the environmental transport processes described above, which affect the intensity and duration of exposures of humans, animals, and plants. With few exceptions, the effects cannot be detected for the levels of radioactivity observed or expected in the environment. Therefore, the effects must be postulated by models that extrapolate the probability of damage to dose levels lower than those used experimentally.

Research on transport systems and ecology is directed mainly toward the assessment of human risk associated with a variety of technologies or specific situations that involve the potential release of radionuclides. Nuclear power production and radioactive waste management are examples of such technologies.

Information gained from environmental research can also be used for reducing radiation risk. Such data are important for establishing standards and regulations, which ensure acceptable risk under normal conditions, and for monitoring compliance with them. In those instances when ecosystem management can delay, divert, or eliminate transport of radionuclides, the prevention or reduction of potential biological effects can be achieved. Environmental research can also provide basic, generally useful information on transport mechanisms and ecological processes. This knowledge can be applied to risk assessment and reduction associated with other pollutants.

Source Factors

Radionuclides are emitted into the air or discharged into waters from numerous sources located throughout the United States. These include nuclear power plants, other facilities involved with the nuclear fuel cycle, national defense facilities, industrial plants, research and development laboratories (e.g., research reactors and accelerators), medical facilities, certain mining and milling operations, and fossil fuel combustion plants. To determine the availability of the released radionuclides to people, biota, and ecological systems, one must obtain the following information for each facility or source category:

- the amounts of radionuclides released, including their release rates, and
- various physical and chemical properties of the radionuclides and concurrently released material, which can influence deposition, removal, resuspension, and uptake of the radionuclides.

Assessments of source factors for both specific locations and generic sites representing many locations are an essential step in the assessment of risk. However, because the FREIR Committee has considered these assessments largely as surveillance rather than research, they have been excluded from this study.

Dispersal Characteristics

To estimate the exposure of humans, biota, and ecological systems to emissions of radionuclides, one must determine the distribution of the radionuclides in the environment. Distribution is influenced by such site characteristics as the release height, plume rise dispersion, and deposition and resuspension due to atmospheric and hydrological factors at the site and along the pathways. The FREIR Committee did not classify the collection of these data as research per se, but did view as research the development and evaluation of predictive models of transport and dose assessment.

Biotic Processes

The consequences of releasing radioactive materials into the environment depend on how the flow of such releases are regulated by either natural or agricultural ecosystems (Auerbach, 1965). The materials may be dispersed and diluted, or they may be concentrated. Early in the history of research on transport, it was learned that certain radionuclides may become concentrated in various components of natural food chains, a phenomenon called bioaccumulation (Davis and Foster, 1958). Biomagnification refers to the increase in bioaccumulation at successive trophic levels. Although this process is not universal, it does have important implications for selected human populations (Hanson and Eberhardt, 1969; Miettinen, 1969). Research contributions made during the 1960's provided new insights into the processes and properties of managed (agricultural) (Benson and Sparrow, 1971) and unmanaged ecological systems (Nelson, 1971).

Physiological Mechanisms

The ultimate health and other biological effects of environmental radionuclides depend on their rates of uptake, retention, and translocation to critical organs. The effects depend on a multitude of physiological processes after ingestion or inhalation. The mechanisms involved, as well as the effects, are discussed in Chapters 5 and 6 of this report.

Predictive Models of Radionuclide Transport and Assessment of Ecological Impacts

Systems ecology can be used to elucidate environmental pathways, to identify the most sensitive components of ecological systems, to predict long-term consequences, and to reconstruct prior exposures to radioactivity. During the 1970's, the large body of information on the behavior of radionuclides in terrestrial, aquatic, and agricultural systems was used to formulate predictive mathematical and computer simulation models of radionuclide transport in the environment (Eberhardt and Nakatani, 1969; Kaye and Ball, 1969) and biomagnification of radionuclides and other toxic substances (Reichle *et al.*, 1970).

The principles of the environmental behavior of many radionuclides, especially those with natural isotopes, are reasonably well understood. Thus, predictive modeling for them is reasonably accurate for regulatory purposes. For example, the generic behavior of strontium, cesium, and iodine is relatively well known. Other radionuclides, such as the transuranium elements and technitium-99, are man-made and have no natural isotopes. The environmental chemistry of these elements is most poorly understood, and their behavior in food chains may be a function of the chemical forms present. Thus, predictive modeling for these radionuclides may not be reasonably accurate, especially for long-term behavior.

The data base for certain aspects of food-chain transport is fragmentary. Data pertaining to the transfer of radionuclides into meat intended for consumption by humans are scarce. There is also a paucity of data on sediment-water interactions and their relationships to the subsequent movement of radionculides into edible aquatic plants and animals.

Models to predict environmental transport of radionuclides and radiation dose to humans and other life forms are currently used by industry and regulatory agencies to determine compliance with existing radiation guidelines and regulations and to estimate impacts of proposed nuclear facilities. They are also used for assessing impacts of radiation resulting from the manufacture, use, and disposal of consumer products containing radioactive materials from nonnuclear technologies, such as combustion of coal, and mining and processing of phosphate, which enhance the release of natural radioactivity to the environment. Because these predictive models are used widely, often for regulatory purposes, there is a pressing need not only to analyze the uncertainties associated with their use but also to validate them by comparison with site-specific measurements.

REFERENCES

Auerbach, S. I. 1965. Radionuclide cycling: Current status and future needs. Health Phys. 11:1355-1361.

Benson, D. W., and A. H. Sparrow. 1971. Survival of Food Crops and Livestock in the Event of Nuclear War. Proceedings of a Symposium held at Brookhaven National Laboratory, Upton, Long Island, New York, Sept. 15-18, 1970. Atomic Energy Commission Symposium Series 24. Brookhaven National Laboratory, Long Island, N.Y. 745 pp.

Davis, J. J., and R. F. Foster. 1958. Bioaccumulation of radioisotopes through aquatic food chains. Ecology 39:530-535.

Eberhardt, L. L., and R. E. Nakatani. 1969. Modeling the behavior of radionuclides in some natural environments. Pp. 740-750 in D. J. Nelson and F. C. Evans, eds. Symposium on Radioecology. Proceedings of the Second National Symposium, Ann Arbor, Mich., May 15-17, 1967. U.S. Atomic Energy Commission, Oak Ridge, Tenn. CONF-670503. Clearinghouse for Scientific and Technical Information, Springfield, Va.

Hanson, W. C., and L. L. Eberhardt. 1969. Effective half-times of radionuclides in Alaskan lichens and Eskimos. Pp. 627-634 in D. J. Nelson and F. C. Evans, eds. Symposium on Radioecology. Proceedings of the Second National Symposium, Ann Arbor, Mich., May 15-17, 1967. U.S. Atomic Energy Commission, Oak Ridge, Tenn. CONF-670503. Clearinghouse for Scientific and Technical Information, Springfield, Va.

Kaye, S. V., and S. J. Ball. 1969. Systems analysis of a coupled compartment model for radionuclide transfer in a tropical environment. Pp. 731-739 in D. J. Nelson and F. C. Evans, eds. Symposium on Radioecology. Proceedings of the Second National Symposium, Ann Arbor, Mich., May 15-17, 1967. U.S. Atomic Energy Commission, Oak Ridge, Tenn. CONF-670503. Clearinghouse for Scientific and Technical Information, Springfield, Va.

Miettinen, J. K. 1969. Enrichment of radioactivity by arctic ecosystems in Finnish Lapland. Pp. 23-31 in D. J. Nelson and F. C. Evans, eds. Symposium on Radioecology. Proceedings of the Second National Symposium, Ann Arbor, Mich., May 15-17, 1967. U.S. Atomic Energy Commission, Oak Ridge, Tenn. CONF-670503. Clearinghouse for Scientific and Technical Information, Springfield, Va.

Nelson, D. J., ed. 1971. Radionuclides in Ecosystems. Vols. 1 and 2. Proceedings of the Third National Symposium on Radioecology, May 10-12, 1971, Oak Ridge, Tenn. Sponsored by the U.S. Atomic Energy Commission, Oak Ridge National Laboratory, and the Ecological Society of America. CONF-710501-P1 and P2. National Technical Information Service, Springfield, Va. 1,268 pp.

Reichle, D. E., P. B. Dunaway, and D. J. Nelson. 1970. Turnover and concentration of radionuclides in food chains. Nuclear Safety 11(1):43-55.

8
Radiation Risk Abatement

The general public receives the major portion of its radiation exposure from so-called natural background sources, both cosmic and terrestrial, and from a series of medical sources including radionuclides, x-ray generators, and particle accelerators used for diagnosis and treatment of disease. Currently, the annual dose equivalent received by the U.S. population from both of these sources amounts to approximately 0.2 rem (National Academy of Sciences, 1980).

Subgroups of the population receive additional radiation exposure in the workplace. Affected persons include radiologists, medical technologists, the employees of nuclear industry, and miners of uranium and other materials who work in formations rich in radioactive materials. The average annual dose equivalent resulting from the occupational exposure of radiation workers amounts to approximately 0.4 rem (National Academy of Sciences, 1980).

The dose-equivalent from natural, medical, and other sources of radiation received by the general public and by radiation workers may be expected to contribute 1% or less to the cancer morbidity and mortality statistics for the United States (National Academy of Sciences, 1980). Because this percentage is small, some might be tempted to argue that efforts to reduce radiation dosage levels both in the public at large and in radiation workers are unnecessary. However, many individuals may be exposed to levels several times higher than the averages cited above. Moreover, even those average levels are sufficiently high to require vigilance as the applications of radiation technology in medicine and industry continue to proliferate.

Exposures from Medical Sources

The medical uses of ionizing radiation have increased rapidly over the years, especially in diagnostic procedures. Currently, one-half of the population is examined radiographically each year (U.S. Department of Health, Education, and Welfare, 1975). A substantial portion is also examined with procedures involving radionuclides. The average annual dose equivalent of the general population from medical sources is approximately 0.1 rem, the same as that received from natural background sources (National Academy of Sciences, 1980).

The Bureau of Radiological Health of the Food and Drug Administration, Department of Health and Human Services, is the federal agency primarily responsible for guiding national policy with respect to medical sources of ionizing radiation. It has a major interest in dose reduction and has supported research, both within its own laboratories and within universities, aimed at improving medical radiation technology. The funding resources available to the bureau, however, have been quite limited.

The diagnostic information yielded by a radiological procedure is closely linked to radiation dosage levels. Therefore, great care must be exercised to assure that the diagnostic information yielded by the procedure is not compromised when dosages are reduced. In recent years, scientists have been quite successful in developing technologies for reducing radiation doses without loss of diagnostic information.

Similarly, research in radiation therapy is actively targeted at risk abatement, i.e., reducing the dose to normal tissue as much as possible while providing a tumoricidal dose. As increasing numbers of cancer patients, especially in the younger age-groups, are cured of their disease through radiation treatment, methodology must be further improved to reduce the probability of subsequent development of radiation-induced tumors.

As the uses of ionizing radiation in medicine continue to increase, a biphasic program of research on medical application and dose reduction must be an important component of any research agenda on the biological effects of ionizing radiation. Unless this is fully recognized, doses of ionizing radiation from medical applications could rise to unacceptable levels, substantially higher than those now prevailing.

Exposures from Cosmic and Terrestrial Sources

The dose equivalent from natural background radiation depends on many variables associated with the radiation's origin. For example, radiation from cosmic sources is closely related to altitude. The annual cosmic dose equivalent at mile-high Denver (~55 mrem) is approximately double that received annually at such coastal cities as Boston, New York, and Philadelphia (~29 mrem) (National Council on Radiation Protection and Measurements, 1975). Although percentage changes with altitude are relatively large, absolute values of dose equivalent are sufficiently small that few individuals elect to reside at sea level on the basis of radiation dosage data.

The terrestrial sources of background radiation are radionuclides present in the earth or those that have transferred from the earth to the atmosphere or hydrosphere. Almost all are primordial in origin. Many of them are isotopes of heavy elements belonging to three radioactive series headed by uranium-238, uranium-235, and thorium-232. In ground surveys in the United States, dose rates in air from natural terrestrial radiation have been found to range from 4 to 180 mrad per year (Oakley, 1972). Although this range is broad, nowhere is the absorbed dose rate considered to be sufficiently high to require individuals to change their residence on the basis of geography.

A terrestrial radionuclide of increasing importance to public health is radon-222, a noble gas and a decay product of radium-226 in the uranium-238 series. This gas emanates from the soil and from building materials of terrestrial origin, e.g., stone, bricks, and concrete. It seeps into homes and office buildings and, when ventilation is restricted, may accumulate in concentrations substantially higher than those prevailing outdoors. In response to the recent need to conserve energy in the heating of homes and office buildings, construction methods that sharply restrict ventilation have been introduced. As a result, the control of radon levels is likely to become increasingly important.

Outdoor concentrations of radon-222 range from 20 to more than 1,000 pCi/m^3 at ground level. Indoor levels are only moderately higher when ventilation is not greatly restricted (National Council on Radiation Protection and Measurements, 1975). In contrast, radon concentrations of 50,000 pCi/m^3 or more have been measured in recently constructed homes designed to limit ventilation as far as possible (R. H. Morgan, personal communication, 1981).

The tissues at risk from exposure to radon include the surfaces of the bronchi, segmental bronchioles, and alveolar membranes. These tissues are exposed primarily to radon daughters, e.g., polonium-218, which attach themselves to dust particles and, when inhaled, deposit themselves within the respiratory system at locations influenced by particle size (National Council on Radiation Protection and Measurements, 1975). Radiation exposure is attributed primarily to alpha particles. The epithelium of alveoli receives an estimated dose equivalent of approximately 0.5 rem per year when radon concentrations in air are 1,000 pCi/m^3. The dose equivalent of the segmental bronchioles is approximately 5 times higher (Harley, 1976). Continuing research and surveillance is needed to monitor radon concentrations in homes and other structures. Moreover, methods of dose reduction need to be developed to assure the conservation of heat while simultaneously preventing substantial buildups of radon concentrations in the ambient air. Sealing techniques, which prevent radon seepage through basement floors and walls, must be an important component of any program to reduce risk from this source.

Occupational Exposure

In most instances, radiation exposure in the workplace has been reasonably well controlled. A notable exception is the uranium mining industry, where it has been especially difficult to maintain ambient radiation levels within acceptable limits. Because of high exposure in the past, the incidence of lung cancer in uranium miners is elevated (Archer *et al.*, 1976). Amelioration of the problem has been difficult because public health authority over the industry has been divided among regulatory agencies. Furthermore, financial support of research to develop improved methods of radiation control in the mines has been limited.

Among an estimated 16,000 people in the United States who have been employed in operations that could involve exposure to plutonium, approximately 5,000 have some measurable evidence of internal plutonium deposition (Barr and Sinclair, personal communication to C. W. Mays, August 1976). Until recently, the U.S. Department of Energy supported research at direct costs of about $500,000 per year to increase the removal of plutonium and transplutonium elements from the body by means such as chelation therapy. In the summer of 1979, however, this research was terminated due to a congressional cutoff in funds (Mays *et al.*, 1981). At the present time, very little research on chelation therapy is being done within the United States. Especially in view of the possibility that breeder reactors and plutonium fuel-reprocessing operations may expose many additional workers to hazards from the accidental intake of plutonium, it would seem appropriate to provide sufficient funding for an adequate research program on chelation therapy.

References

Archer, V. E., J. D. Gillam, and J. K. Wagoner. 1976. Respiratory disease mortality among uranium miners. Pp. 280-295 in U. Saffrotti and J. K. Wagoner, eds. Occupational Carcinogenesis. Vol. 271. Ann. N.Y. Acad. Sci., New York.

Harley, J. H. 1976. Environmental radon. Pp. 109-114 in R. E. Stanley and A. A. Moghissi, eds. Noble Gases. (Based on a symposium held in Las Vegas, Nevada, Sept. 24-28, 1973, cosponsored by the U.S. Environmental Protection Agency's National Environmental Research Center at Las Vegas and the University of Nevada, Las Vegas.) EPA No. 600/9-76-026. National Technical Information Service, Springfield, Va.

Mays, C. W., G. N. Taylor, R. D. Lloyd, and M. E. Wrenn. 1981. Status of chelation therapy: A review. Pp. 351-368 in M. E. Wrenn, ed. Actinides in Man and Animals. Proceedings of a workshop held in October 1979 at Snowbird, Utah. Radiobiology Division Press, University of Utah, Salt Lake City.

National Academy of Sciences. 1980. The Effects on Populations of Exposure to Low Levels of Ionizing Radiation. Report of the Committee on the Biological Effects of Ionizing Radiations. National Academy of Sciences, Washington, D.C. 638 pp.

National Council on Radiation Protection and Measurements. 1975. Natural Background Radiation in the United States. NCRP Report 45. National Council on Radiation Protection and Measurements, Washington, D. C. 163 pp.

Oakley, D. T. 1972. Natural Radiation Exposure in the United States. Report ORP/SID-72-1. U.S. Environmental Protection Agency, Office of Radiation Programs, Washington, D.C.

U.S. Department of Health, Education, and Welfare. 1975. Population Exposure to X-rays, U.S., 1970. Public Health Service, Food and Drug Administration, Bureau of Radiological Health. Publ. DHEW/PHS/FDA/BRH 73-8047. National Technical Information Service, Springfield, Va. 137 pp.

9
Scope and Quality of Current Research

Research on the biological and health effects of ionizing radiation is supported by 15 federal agencies. Each agency conducts its work relatively independently of the others and designs its research programs to meet its specific objectives. Only approximate figures of federal expenditures for this research are presented because of different accounting procedures among and within the federal agencies and because a number of research projects include items that do not appear to be related to the mission of this committee.

Table 9-1 summarizes the federal support of research on the biological effects of ionizing radiation as provided by agency sources during the budget years 1979-1981. The data indicate that approximately 50% of the research has been and will be supported by the Department of Energy (DOE). The National Institutes of Health (NIH), The Department of Defense (DOD), and the Department of Agriculture (DOA) also provide substantial research support. The figures in this table, when converted to constant dollars, indicate that this research support has been declining in recent years.

This chapter summarizes the committee's evaluation of federally supported research on the biological effects of ionizing radiation currently in progress in the United States. This research is but a small portion of an investigative effort that had its beginnings more than 40 yr ago. In many ways, today's research represents the fulfillment of planning that also had its origins in earlier times.

TABLE 9-1 Support of Research on the Effects of Ionizing Radiation[a]

		Research Support by Category ($1,000)				
Year	Agency[b]	Biological	Ecological	Human Health	Other	Total
1979	DOE	21,000	10,600	19,200	—	50,800
	NIH	10,800	100	6,400	—	17,300
	DOD	7,500	—	2,800	1,000	11,300
	USDA	9,500	—	—	—	9,500
	FDA/BRH[c]	1,400	—	600	—	2,000
	NSF	—	—	—	2,000	2,000
	NRC	1,000	200	400	—	1,600
	NBS	—	—	—	1,000	1,000
	EPA	200	700	—	—	900
	NIOSH	—	—	800	—	800
	NASA	—	—	800	—	800
	CDC	—	—	—	500	500
	VA	200	—	100	—	300
	NCHS	—	—	200	—	200
	DOT	—	—	—	100	100
Total Federal Support		51,600	11,600	31,300	4,600	99,100
1980	DOE	21,500	10,000	20,000	—	51,500
	NIH	13,200	100	7,600	—	20,900
	DOD (R&E)	8,000	—	3,200	1,400	12,600
	USDA	10,400	—	—	—	10,400
	FDA/BRH	1,700	—	700	—	2,400
	NSF	—	—	—	2,000	2,000
	EPA	200	1,400	100	—	1,700
	NRC	700	200	500	—	1,400
	NBS	—	—	—	1,000	1,000
	NIOSH	—	—	800	—	800
	NASA	—	—	800	—	800
	VA	200	—	100	—	300
	NCHS	—	—	200	—	200
	DOT	—	—	—	100	100
	CDC	—	—	—	100	100
Total Federal Support		55,900	11,700	34,000	4,600	106,200
1981	DOE	23,200	10,700	21,400	—	55,300
	NIH	14,200	100	8,000	—	22,300
	DOD	8,100	—	1,800	1,700	11,600
	USDA	10,800	—	—	—	10,800

TABLE 9-1 (Continued)

		Research Support by Category ($1,000)				
Year	Agency[b]	Biological	Ecological	Human Health	Other	Total
	FDA/BRH	1,700	—	700	—	2,400
	EPA	500	1,500	300	—	2,300
	NSF	—	—	—	2,000	2,000
	NRC	1,000	200	600	—	1,800
	NBS	—	—	—	1,100	1,100
	NASA	—	—	900	—	900
	NIOSH	—	—	400	—	400
	VA	200	—	100	—	300
	NCHS	—	—	200	—	200
	DOT	—	—	—	100	100
	CDC	—	—	—	100	100
Total Federal Support		59,700	12,500	34,400	5,000	111,600

[a]Unpublished data from Interagency Committee on Federal Research into the Biological Effects of Ionizing Radiation, not to be compared with later compilations since the figures are agency estimates as of September 1980.

[b]DOE, Department of Energy
NIH, National Institutes of Health
USDA, U.S. Department of Agriculture
FDA/BRH, Food and Drug Administration, Bureau of Radiological Health
NSF, National Science Foundation
NRC, Nuclear Regulatory Commission
NBS, National Bureau of Standards
EPA, Environmental Protection Agency
NIOSH, National Institute for Occupational Safety and Health
NASA, National Aeronautics and Space Administration
CDC, Centers for Disease Control
VA, Veterans Administration
NCHS, National Center for Health Statistics
DOT, Department of Transportation

[c]Excludes all agency overhead.

As would be expected, changes in emphasis have taken place in the intervening years. Nonetheless, the current investigative effort clearly bears the mark of plans laid out shortly after World War II. The scope of today's research retains much of the cohesiveness of earlier days despite recent fragmentation of the federal agencies that support research on the biological effects of radiation.

The committee finds the quality of current research to be generally good. With few exceptions, it appears to be well conceived and carefully pursued by competent scientists. This is attributable largely to the procedures used by federal agencies to determine that the research objectives and experimental designs of work proposed by their contractors and grantees are appropriate and that the work is carefully and diligently conducted after it has begun. These procedures differ from agency to agency. Some, such as the NIH, use a process of external peer review in which research proposals within a given discipline are reviewed by scientists of a similar discipline. In other agencies, such as the DOD or Department of Transportation (DOT), research proposals may be evaluated by agency staff members knowledgeable in the proposals' subject matter with external peer reviewers for "off-site" laboratories. Variations and combinations of these procedures are also used. Each of the systems has its advantages and disadvantages. However, with the increasing number of federal agencies involved in radiation research, the committee believes that there is merit in the adoption of systems of evaluation and review that are comparable to each other. Of the various systems in use, external peer review seems to be the most objective and provides a means of introducing a broad range of expert scientific guidance to the evaluation and review processes.

The remainder of this chapter is divided into nine topics:

- Dosimetry
- Effects in humans
- Mutagenesis in bacteria, plants, lower animals, and mammals
- Effects at cellular and molecular levels
- External and internal radiation in animals
- Environmental exposures, radionuclide transport, and effects
- Occupational exposures
- Reduction of radiation risk from therapeutic irradiation
- Reduction of radiation risk from diagnostic radiation

Dosimetry

Dosimetry is the branch of the radiation sciences concerned with the measurement of certain physical quantities associated with ionizing radiation, including absorbed dose, dose equivalent, exposure, and radioactivity. As discussed in Chapters 3 through 6, the ability to measure these and related quantities under a wide range of circumstances is an essential element of all research pertaining to the biological and health effects of ionizing radiation. This ability is also an essential element of all programs to control and regulate the exposures of humans to ionizing radiation.

Radiation monitoring instruments may be divided into two general classes: those for measuring radiation fields and those for measuring radioactivity. Several types of radiation detectors fall into both classes.

These include ionization chambers, proportional counters, absorbed dose calorimetry, scintillation detectors, semiconductors, Cerenkov counters, films, and radiophotoluminescents (National Council on Radiation Protection and Measurements, 1978a,b).

The instrumentation usually includes a detector as well as electronic pulse measuring and recording equipment. Instruments range from simple systems for routine work to very elaborate systems that include capabilities such as coincidence-anticoincidence counting, Compton radiation suppression, and compound detectors for special purposes. These complicated systems are used when there is a need to measure not only the total absorbed dose but also temporal, spatial, and energy distributions of the doses.

At present, the general dosimetric requirements are being met for most types of ionizing radiation, including x-rays and gamma rays deposited in relatively large volumes of material. Difficulties arise when dealing with such radiations as neutrons, pions, or heavy ions, with mixed-radiation fields, with radiation emitted by radionuclides deposited within the body, or with radiation absorbed in volumes as small as, or smaller than, a human cell. A strong and stable effort in research on dosimetry, including the improvement of existing equipment and facilities and development of new equipment to meet foreseeable research needs, is essential if both regulatory and scientific goals are to be met.

Automation of radiation measurements is less desirable when more detailed measurements are required for scientific purposes since they tend to require longer counting times. These dosimetric requirements are not well met. One reason is the lack of variety among the types of detectors that are available (Harley, 1980). For example, both alpha and beta scintillation counters offer high efficiency and low background, but no such instruments are listed in the manufacturers' catalogues (Harley, 1980).

Efforts are being devoted to improving the quality of measurements, expediting the counting, and simplifying the techniques. The major breakthroughs are most likely to come from the basic scientific community. As in all fields, production of less expensive systems would allow wider use and assist in evaluating the exposure of humans. Transfer of instrumentation technology among the major laboratories and to the commercial sector has been exasperatingly slow, and there is no apparent mechanism for improvement (Harley, 1980).

More sophisticated techniques would be particularly useful for monitoring in and around facilities where there are mixed-radiation fields, especially those containing fast neutrons. Current dosimetric techniques are not sufficient for measuring fast neutrons because of their complex interactions with matter, the broad mix of secondary particles that are produced, the diversity of the energy density, and relative biological effectiveness (RBE) of both the neutrons and their secondary particles. The published neutron and gamma-ray doses to the atomic bomb survivors have recently been challenged by physicists at the Lawrence Livermore Laboratory. Because current estimates on the risks from radiation-induced cancer depends so heavily on the dose-effect relationship in the atomic bomb survivors, the uncertainty in tissue doses needs to be resolved.

Beams of energetic heavy ions from particle accelerators are finding increased application in medical diagnosis and therapy. Conventional dosimetric measurements are not sufficient for these radiations because they fail to provide sufficient information to describe the radiation resulting from nuclear fragmentation of the heavy nuclei, a process that produces a spectrum of particles with low atomic numbers. Moreover, they do not provide an adequate indication of the spatial distribution of the dose adjacent to the targeted tissue or of the characteristics of energy transfer necessary for the derivation of RBE of the mixed radiation.

Additional work to improve dosimetric techniques must be pursued in order to derive RBE values. Small-volume dosimetry, which deals with the energy that is deposited in very small quantities (cubic millimeters), can provide information needed in radiobiological studies, especially in those pertaining to the mechanisms of radiation effects. Microdosimetry, which is the measurement of radiation in even smaller volumes (cubic micrometers), attempts to provide the information needed in this area.

The deposition of radionuclides within the body is not uniform due to effects exerted by both physical and biological processes. Thus, the dosimetry of internal radionuclides is difficult. Whole-body counting and bioassays indicate the amount of radioactivity present, but not its distribution. At present, information on the distribution of radionuclides is obtained mainly by postmortem studies. Improvements in *in-vivo* measuring are needed to facilitate risk assessment, to reduce radiation exposures, and to meet the requirements of various research programs.

Research Objectives

Research on dosimetry is needed to provide answers to the following questions:

• Is dosimetric technology adequate for regulating and controlling exposure of humans to ionizing radiation? If not, what are the shortcomings?

• Are dosimetric technology and procedures adequate for conducting the basic research programs that are necessary to increase our knowledge of the biological effects of ionizing radiation? If not, what are the most promising ways to improve them?

• How can dosimetric technology and procedures be simplified and made more cost-effective than they are now?

In order to answer these questions, the FREIR Committee reviewed the literature, held a workshop, interviewed investigators with expertise in specific research areas, and examined in depth 21 out of 50 research projects identified in this area. Some of them were purely dosimetric research projects; many others had dosimetric components.

Most of the research in dosimetry is sponsored by DOE. Few studies are funded by the National Bureau of Standards (NBS), NIH, the National Aeronautics and Space Administration (NASA), and the Bureau of Radiological Health (BRH). Two of the projects reviewed by the committee were sponsored by NIH; the other 19 were funded by DOE.

Quality of the Research

The committee concluded that the projects reviewed were productive and of good quality and that the instrumentation has been developed to meet identified needs. The exceptional quality of this research may have been influenced by the fact that declining support has generally resulted in a decline of output but in more competition and, generally, the survival of only the better projects.

Adequacy

The committee believes that all but one of the projects reviewed were important and relevant. The dosimetric requirements for regulating and controlling exposure of humans to electromagnetic radiation are generally being met; however, greater precision is needed in the dosimetry of high-energy x-rays, neutrons, and mixed-radiation fields. Furthermore, new technologies are required in small-volume dosimetry and in *in-vivo* measurements of radionuclides deposited within the body.

Recommendations

• The committee concluded that dosimetric capabilities are reasonably adequate for electromagnetic radiation and charged particles, but that dosimetry for neutrons and for mixed-radiation fields is in an early stage of development and requires more research. Uncertainties in the doses to the atomic bomb survivors should be resolved, especially for neutrons.

• *In-vivo* measurements to determine doses from nonuniform distribution of radionuclides deposited within the body should also be improved.

• The assessment of radionuclide doses to particular organs and to specific cells of organs is an area of continued need. Studies of radionuclide uptake, deposition, metabolism, and elimination play an important role in health protection and should therefore be encouraged. Priority should be given to radionuclides to which large populations of humans may become exposed.

Effects in Humans

During the past 30 yr, research programs have produced a substantial volume of knowledge about the biological effects of ionizing radiations. These investigations have been of varying scientific quality and varying relevance to objectives, but much of it is judged to be good or excellent.

Studies of human populations exposed to ionizing radiation, e.g., those exposed to atomic bombs, have contributed in important ways to our knowledge of radiation effects and the associated risk for humans. Yet, the ability to predict risks is hampered by too little understanding of biological effects at the low doses received by most persons and by the lack of sensitive indicators of effects in humans.

The committee reviewed 24 studies concerning the effects of ionizing radiation on human populations. This category included epidemiologic studies relating to occupational, diagnostic, accidental, military (including atomic bomb exposures), therapeutic, and fallout exposures from ionizing radiation; studies of human cells taken from exposed persons; biostatistical manipulation of exposure data; core and support grants allocating money specifically for epidemiologic

TABLE 9-2 Distribution Among Agencies of Epidemiological Studies Reviewed by Committee

Agency	Submitted Studies	Reviewed Studies
DOD	3	1
DOE[a]	24	3
EPA	1	1
CDC	3	1
NIOSH	2	1
BRH	9	3
NCHS	3	3
NIH		
NCI	26	6
Other	6	
NRC	4	2
VA	7	3
TOTAL	88	24

[a]Fourteen of these were reviewed by the Committee on DOE Research on Health Effects of Ionizing Radiation (National Academy of Sciences, 1980a).

radiation research; and registries, surveys, data collection, and storage programs based on exposed populations.

Lists totalling 149 studies were received from 11 federal agencies. A substantial number of these were clinical trials, i.e., studies to determine the benefits of radiation in therapeutic regimens. Such studies were not designed to provide data on risks. In several other studies, radiation was used as an incidental technique. Another group of studies had been terminated prior to the date selected for review of the projects to be included in the present survey. Deletion of the clinical trials and the irrelevant and terminated studies left 88 epidemiologic studies appearing to fall within the purview of this committee (Table 9-2).

Research Objectives

Research on irradiated persons or human populations has as its objective the measurement of the health effects caused by radiation and the prediction of such effects for as yet unexposed persons. Populations include atomic bomb survivors; those exposed during military exercises; persons occupationally exposed, such as uranium miners, naval shipyard employees, and nuclear plant workers; persons receiving therapeutic radiation; and persons living in areas with high background radiation. Both external irradiation and internally deposited emitters are under study.

Both somatic (primarily cancer) and genetic effects are being studied. A substantial portion of this effort is directed toward developing more effective methods for measuring genetic effects. Among the variables under investigation are the kind of radiation, acute versus chronic exposure, effects at different intensities of radiation, the influence of age at time of exposure, and the risk over time after exposure.

Quality of Research

As might be expected, the studies vary in quality, but most of the research appears to be conducted competently. The poorer quality of some of the studies generally was associated with inadequate review of research proposals by the agencies, and often, the lack of trained epidemiologists involved with the projects was a major factor.

Adequacy of Research

One charge to the committee was to determine if the funded research was adequate to meet the objectives of the various federal agencies.

The committee's review indicates there is no overall plan that encompasses all agencies. Therefore, the approach to research support is fragmented. This haphazard system has the potential to result in omissions of some research and duplication of other efforts. However, only a few important examples of such deficiencies (e.g., uranium miners) or duplications (e.g., shipyard workers) were found.

Epidemiologic studies of exposed human populations to detect harmful effects of ionizing radiation require long-term support. Since the appearance of some solid tumors is often delayed for decades, individuals must be observed over long periods. Therefore, human studies are generally of much longer duration than those conducted on animals.

The committee found a notable failure of the federal agencies to provide adequate and continuous support for one of the most important radiation studies—that of U.S. uranium miners. The federal government does not now have a mechanism for the full utilization of the epidemiologic data that are being gathered at great cost. It would be valuable to devise a system to make such data available to subsequent researchers who might wish either to extend them or to analyze them more fully.

Paramount among the factors to be considered when evaluating the adequacy of proposed research is the degree to which the studies will contribute to the scientific priorities discussed in Chapter 4. Briefly stated, they must fall into one or more of the following categories:

- studies with a potential to increase our knowledge about the shapes and slopes of the dose-response curves;
- studies that provide information on inadequately characterized end points, e.g., germinal mutations;
- studies that provide needed information on the temporal patterns of radiation risk;
- studies that provide information on dose-rate or dose-fractionation effects or on RBE;
- studies that verify or elucidate biochemical mechanisms that facilitate our ability to generalize from animal models; and
- studies that add to our fund of knowledge on the nature and degree of modification of radiation effects by other exposures or by host susceptibility factors.

Proposed studies should also be carefully reviewed for scientific adequacy. They should have adequate dosimetry, an adequate range of doses and/or adequate "signal detection" capability, and adequate control of potential confounding variables.

A wide range of disciplines is required for studies of radiation effects in humans. These include radiation biology, radiation physics, epidemiology, biostatistics, data management science, genetics, clinical medicine, and pathology. The committee did not systematically study the manpower needs in these areas. It did note, however, that there is a lack of manpower in a number of disciplines, notably in the field of epidemiology. Epidemiologic positions in government agencies concerned with effects of ionizing radiation are unfilled. The small group of epidemiologists working in this area is insufficient to respond to the many requests for research assistance and guidance. Interdisciplinary training programs encompassing quantitative epidemiology, radiation biology, and allied areas are badly needed.

Recommendations

Establishment of priorities for research on human health effects of ionizing radiation appears to have been more haphazard than orderly. In recent years, a priority imbalance appears to have developed in the direction of excessive effort and support directed at epidemiologic studies of populations exposed to low radiation doses. The committee recommends that existing studies in this category be carefuly evaluated as to their probability of providing scientifically useful information and that fewer such studies be supported. In the dichotomy between basic research and applied research, some individual scientists argue for reconsideration of priorities favoring one or the other, whereas others find the present balance approximately correct.

Radiation Studies on Mutagenesis in Bacteria, Plants, Lower Animals, and Mammals

Genetic damage induced by radiation can be classified as either gene mutations or mutations resulting from the breakage and rejoining of chromosomes. The work in this field has provided the data base from which radiation-related risks to future generations can be estimated.

Because there are many types of mutations in both genes and chromosomes, the exact nature of the formations is still not well understood. Some of the mutations are simply base changes in DNA, and others are small duplications and deficiencies; whereas the gross chromosomal mutations affect the organization and relative amounts of DNA and its associated protein, which comprise the chromosome. Neither the relative number of mutations that are deletions nor the kinetics of their induction by ionizing radiation are well understood in mammals, partly because the radiation kills germ cells, leading to distortions in the shapes of the dose-response curves. Thus, specific locus mutations in the male mouse increase linearly with dose but are also subject to a dose fractionation effect of the type expected for phenomena that increase at a rate greater than the first power of the dose. Arguments have been presented that these mutations are indeed one-hit phenomena that are subject to repair between dose fractions. A known two-hit phenomenon, the induction of genetic translocations, also increases linearly with dose in the same system, leading to the alternative theory that the specific locus mutations are mainly two-hit deletions and that the dose fractionation effect is caused by the repair of chromosome breaks, which will decrease the numbers of breaks capable of interacting. This view is strengthened by the fact that the dose-response curve for specific locus mutations in the female mouse increases faster than the first power of the dose over the entire dose range tested. However, it has been postulated that this nonlinear curve is really a repair-induced distortion of a linear curve (National Academy of Sciences, 1980b).

Research Objectives

To improve our knowledge and understanding of radiation-induced mutation rates, we need further information on the nature of the various forms of genetic damage, their relative frequencies, their modification by repair mechanisms, their heritability, their transmission, and their biological consequences. The shapes of dose-response curves must be precisely determined to obtain estimates of risk at low doses where it is impracticable to obtain epidemiologic data in humans. To this end, radiation genetic studies at the molecular and cellular level are necessary to enable us to draw reasonable conclusions regarding the validity of extrapolations of dose-effect relationships to low doses from data obtained at higher doses. This information obtained at the fundamental level will then have to be incorporated into information obtained with whole animals. The role of reproductive processes, such as meiosis and gametogenesis in the ultimate genetic response, needs to be elucidated.

Scope and Quality

Investigators for 149 projects supported by DOA; DOE; NASA; the Veterans Administration (VA); the Department of Health, Education, and Welfare (HEW), now the Department of Health and Human Services (DHHS); and the Environmental Protection Agency (EPA) listed themselves as carrying out studies in radiation genetics. Many of these studies involved the use of radiation to induce mutations useful for plant breeding or the use of radiation-induced sterility to control insects. Other studies dealt with effects of radiation on DNA and DNA repair, addressing the effects of high- and low-LET radiations, as well as those of incorporated radionuclides. Of these studies, 14 were reviewed either by site visits or by paper reviews. As in any area of science, some of the work was excellent and some only mediocre. In general, however, most of the work was of very high quality and addressed important issues.

Studies in progress are examining the induction of specific locus mutations, their modification by different dose regimens, and their nature. The role of DNA synthesis and repair in the induction of specific locus mutations and frank chromosome aberrations is being elucidated, as are the types of heritable genetic damage induced by densely ionizing radiation and incorporated radioisotopes. Attempts are being made to estimate the genetic risk of radiation in various organisms and then to extrapolate the results to estimate the risk to humans. Much fundamental work is being done in both genes and chromosomes to determine the nature of the mutations and their repair, as well as to gain basic knowledge of genetic mechanisms. Ongoing studies in microorganisms, insects, plants, animals, and cells in culture, including human cells, will ultimately be extended to somatic cells and germ cells.

In the past, radiation was the mutagen of choice in genetic research. Today, many chemical mutagens are known and there is an increasing awareness of environmental risks other than those of radiation. Thus, many investigators who might have used radiation in the past are selecting alternative mutagens for their studies. As a consequence, progress in radiation genetics is slower than it might otherwise have been. In this sense, the radiation-specific work is not proceeding at an optimal pace, although work at the fundamental level is proceeding well.

Recommendations

Studies should be conducted to obtain information that will assist in the determination of the nature of radiation-induced mutations, their dependence on repair mechanisms, and the influence of the type of radiation on risk estimates (e.g., LET).

Effects at Cellular and Molecular Levels

This material is based on the review of reports on 23 projects and the results of site visits to three laboratories. This research is concerned with the actions of radiations in biological systems at the cellular and molecular levels and is undertaken to seek answers to the following questions:

- What are the mechanisms of energy absorption and dissipation and the reactions of excited molecules and free radicals with the macromolecular structures making up cells?

• How do the workings of biological systems influence the effects of radiation? For example, how essential is the integrity of DNA for cell survival and what are the parameters controlling cell transformation?

• Can the conclusions learned from radiation biology with its excellent dosimetry be carried over to other potentially deleterious environmental agents?

• Radiation damages were the original models for DNA damage and repair. How general and what are the characteristics of DNA repair systems? What are the genetic or other controlling mechanisms for such repair systems?

• Can theoretical models be constructed to explain the reactions of cells to radiations in terms of physical dosimetric concepts and damages at the macromolecular or microstructural levels?

• Can one combine the results of analysis of radiation effects at the molecular, cellular, and whole-animal levels with appropriate theories to predict with confidence the genetic and carcinogenic effects on humans at low doses or at high doses accumulated over long periods?

Research Objectives

The specific objectives of the projects studied are aimed at answering the general questions posed above. The methods used attempt to correlate dosimetric measurements and molecular changes with the shapes of survival curves, the dose-response curves for transformation, and measurements of DNA repair *in vivo* and *in vitro*. A number of studies are attempting to explain the cancer proneness of certain human diseases in terms of repair efficiencies in cells of the affected individuals. Such repair deficiencies are inferred either from survival as a function of dose or from direct measurements of macromolecular repair in such cells.

Quality of Research

The studies reviewed covered the effects of radiation on small molecules, DNA, the proliferation of cells, and the transformation of cells; the manner in which such effects are influenced by the repair capabilities of cells; and the relationship of the effects to the cell cycle and macromolecular syntheses. A number of studies were concerned not only with the effects of ionizing radiation, but also with the effects of ultraviolet radiation (UV) and the effects of chemical carcinogens on human and other mammalian cells. Such studies not only attempt to generalize the effects of chemical and physical carcinogens, but also utilize knowledge about the specific products formed by UV or chemical carcinogens in attempts to elucidate the mechanisms of DNA repair *in vivo* and *in vitro*. Thus, there are strong efforts to purify repair enzymes from a number of mammalian and bacterial sysems. In some of these studies, investigators are trying to gain an understanding of the molecular nature of the lesions induced by ionizing radiation. So far, there has been no unique correlation of such lesions with biological effects, nor is it apparent what dose rates will saturate the repair system for ionizing radiation damage.

A general feature of all the projects examined by the committee is their high quality. Their results are published in the refereed literature, and the projects themselves are reviewed regularly. The investigators, with only one or two exceptions, know clearly what they are attempting to do and the limitations of the methods they are using. Almost all of the investigators would probably use better methods if any were available. Rarely, however, did any one project—even the longer ones—encompass extensive study of both biological and macromolecular effects of ionizing radiation. This deficiency arises because the important molecular lesions have not been clearly identified and studies are often dictated by ease of experimental observation rather than by their relevance. Theories are being developed to relate radiation dose and survival as a function of time in the cell cycle at irradiation, the effects of radiation on macromolecular synthesis, and the alteration of cellular responses by DNA repair. Also under investigation are the DNA repair systems for ionizing-radiation-induced damage, such as single-strand nicks, double-strand breaks, and several types of base damage. A number of theories have been put forward to explain the observed phenomena. In theory, they can be tested experimentally; however, they have not yet been subjected to crucial testing. The success of such testing appears to await a firmer understanding of the biologically important damages introduced into DNA by ionizing radiation at the macromolecular level and the mechanisms through which these damages are biologically effective. Investigators in this area are convinced that future advances will depend not only on identifying the molecular nature of the lesions involved, but also on a detailed knowledge of cell biology, i.e., the molecular basis for controlling the function of cells.

Scope and Adequacy

Both DOE and NIH, which provided funding for the projects reviewed, appreciate that extrapolation of the data from high to low doses and dose rates involves many uncertainties, not only for ionizing radiation, but also for other natural or man-made pollutants, and that knowledge about the molecular workings of human cell systems is gained not only from experiments in human cells, but also from those in other eucaryotic systems, procaryotic cells, and virus-infected cells. Thus, the overall scope of the supported research seems adequate, but the distribution of financial support may not be. It is much easier to obtain money for studies on the effects of carcinogenic agents on human cells than for those on the effects of mutagenic agents on bacteria, despite the fact that in the past more information on molecular effects and various repair processes was gained from the latter systems. However, even these effects are not well understood. It is true that bacteria do not get cancer and that the DNA in human cells is organized differently than it is in bacterial cells. Nevertheless, extrapolations from molecular to cellular effects at the single-cell level require a detailed knowledge of molecular damages of DNA, their lifetimes, their repair, and the effects of such damages on replication, transcription, and translation. It is difficult to see how an extrapolation theory can be worked out for human eucaryotic cells, when it is not adequately understood for procaryotic bacterial cells. The latter point of view seems to be well understood in the world of chemical carcinogenesis and mutagenesis and may be well understood for ionizing radiation, but it certainly does not seem to be implemented effectively.

Recommendation

Research funding should be expanded to include not only the effects of carcinogenic and mutagenic agents on human cells, but also those on animal cells, plants, and bacteria.

Studies of External and Internal Radiation in Animals

For many radiation dose-exposure patterns, namely total dose, dose rate, fractionation, protraction, and LET, there is insufficient data for humans. However, these factors do apply to the exposure of humans to ionizing radiation. Animals can serve as surrogates for humans in studies to develop dose-response data and to elucidate mechanisms of responses to radiation exposure. Such experiments in animals supplement and extend the data on humans in one or more of the following three areas:

- risk assessment of internal and external radiation exposures and collection of information pertaining to occupational and environmental exposure standards;
- determination of how the various physical, physiological, and environmental factors modify radiation dose-response relationships; and
- identification and definition of the mechanisms that act in whole animals to suppress or promote the eventual expression of ionization events in the cell nucleus as genetic defects or neoplasia.

Twelve of 23 studies on externally irradiated animals were reviewed by the committee. Two projects were supported by NASA, one by the Nuclear Regulatory Commission (NRC), ten by agencies of DHHS, and nine by DOE. In these studies, animals were exposed to graded doses of external radiation to define dose-response relationships and to determine the influence of various physical, biological, and environmental factors on these relationships.

Seventy projects involving external or internal exposures to whole animals were not reviewed. The committee had decided that these studies had minor relevance to the understanding of the long-term effects on humans resulting from exposure to radiation. Forty-five of these projects were sponsored by DOD and dealt mainly with the acute effects of radiation on various organs and tissues under combat conditions. Most of the other 25 projects were supported by NIH. In these, radiation was used mainly as a tool to support clinical radiology, to suppress the immune response, or to probe the physiology or biochemistry of specific tissues and organs.

Twenty-three of 30 identified studies on internally irradiated animals were reviewed in depth. Twenty-seven of these were radiotoxicological studies in which graded dosages were administered to animals to obtain quantitative dose-response data. The other three projects provided background dosimetry and physiological data to support the others. Twenty-five projects were supported by DOE, four by DHHS, and one was sponsored jointly by DOE and EPA.

Objectives

Studies of externally irradiated intact animals address mainly the following scientific questions:

- What are the shapes of dose-response curves for gamma rays in the region below 100 rad for neoplasia and for life-shortening?
- How are the shapes of dose-response curves modified by physical factors such as dose rate, dose fractionation, and dose protraction and by physiological factors such as sex, age at irradiation, and the fraction of the life span over which the radiation is delivered?
- What are the shapes of dose-response curves for fission neutrons? Is there a low-dose region for neutrons where enhancement of effects of fractionated doses ceases and simple additivity or repair can be demonstrated?
- How are the latent periods for various animal neoplasias changed by total dose, dose rate, dose fractionation, dose protraction, radiation quality, or stage of development of the animal? Do latent periods for tumors of the same tissues differ among species or among the various tumors in the same animal?
- Are there interspecific differences in the manifestation of radiation effects, e.g, in the incidences of different tumors. If so, are such differences due to the physical aspects of the radiation or to physiological factors, such as life span, body size, or genetic constitution?
- Can new animal tumor models be developed in which tumor yield per rad can be increased by "promotion," thereby permitting definition of the shapes of dose-response curves at lower doses and dose rates?

The major objectives of animal studies on the effects of internally deposited radionuclides are evaluation of the effects and provision of dose-response data for radionuclides deposited in various tissues or organs and modification of tissue doses and responses by physiological and environmental variables.

Scope

Constraints on facilities, staffs, and budgets have tended to reduce the number of studies of irradiation of whole animals and to limit their scope. Such studies are now confined largely to specialized laboratories. The only species now under intense investigation are mice, rats, and beagle dogs, and few radiation sources are used, except energetic gamma rays and fission spectrum neutrons. Considerable emphasis is placed on fractionated or protracted exposures. Although these exposure patterns are of great practical and theoretical interest, they cannot yield good replications in laboratory animals of the single, acute exposures of the populations that provide the largest source of data on radiation risk in humans—the atomic bomb survivors.

Most opportunities for observation of late effects in large animals have been lost, because animals surviving acute lethality studies could not be supported and were killed prematurely. The one exception is a small herd of burros irradiated at the Nevada Test Site with A-bomb gamma rays and fallout nuclides. These animals are being held for life-span observation.

There is no adequate and completed dose-response study of late effects resulting from single exposures to gamma rays in any large, relatively long-lived mammal. Several groups of young adult beagle dogs have been irradiated with x-rays or gamma rays at different times in different laboratories. However, even when those data are combined, they do not define the dose-response curves for either life-shortening or induction of neoplasms at doses below 100 rad.

There is a great interest in the late effects of neutron irradiation. High-energy neutrons are being studied to determine their usefulness in radiation therapy.

Although the scope of the projects on inhalation exposures is broad, there are no toxicity studies on chronic inhalation of thorium-232 and thorium-228, which would be important if the U.S. government chooses to construct a thorium-fueled nuclear reactor. Past studies on that subject were terminated before enough data had been accumulated. Also terminated was research on removal of radionuclides from the body. These studies, especially those on the removal of insoluble particles from the lung, should also be reinstated, especially if policies on nuclear fuel reprocessing and the development of the breeder reactor become accepted.

Quality

The scientific staffs of the mammalian radiobiology projects that were reviewed are well trained. Most of the investigators were neither originally nor exclusively trained in radiation biology. Rather, they had obtained advanced degrees in such other scientific specialities as medicine, veterinary medicine, veterinary pathology, physics, chemistry, biochemistry, or immunology. Most of the senior investigators in the long-term studies participated in the design of the current experiments. Consequently, they have considerable personal interest from the standpoint of their careers as well as scientific interest in the results, and they are dedicated to seeing the projects through to completion.

There are problems common to all of the long-term projects in which large numbers of pathological specimens must be examined to produce data on causes of death and incidence of neoplasia. There is either insufficient technical support to prepare specimens or too few trained pathologists to examine them. This insufficiency is often a serious impediment to the timely completion of long-term projects on late radiation effects.

The physical facilities, e.g., laboratories; animal breeding, exposure, and holding facilities; radioactivity measuring equipment; and computer and data storage systems, are generally well designed and well maintained. The support staffs appear to be well trained, if not always adequate in numbers for greatest effectiveness.

There is emphasis on use of high-quality animals, whether they are raised on the premises or purchased from dealers, and on good animal care practices.

Timely reporting of long-term studies in peer-reviewed journals is a constant problem. Although prompt reporting of results is encouraged in all projects reviewed by the committee, the investigators frequently choose to wait until an experiment is complete before preparing a final paper. However, there are mechanisms for interim reporting, such as annual progress reports and budget documents submitted to the sponsoring agency, abstracts and oral presentations at professional society meetings, and interim reports prepared for international meetings and topical symposia. All of these are used to varying degrees by the various projects.

Responsiveness

Site visit presentations and written objectives submitted for review indicate that investigators are acutely aware of the criteria for agency funding of studies in mammalian radiobiology. To be accepted, proposals submitted to an agency staff or to a group of anonymous peer reviewers must fill a definable need for information to support risk estimation and the establishment of protection standards or to improve medical diagnosis or treatment.

Nine large-scale and/or long-term animal studies are based in national laboratories, specially constructed facilities located at universities, and a medical institute. Many of the experiments that constitute these large projects are the end products of extensive consultations among the investigators, program and laboratory directors, the staffs of the sponsoring agency, and consultants to the agencies. Thus, the agencies have significant responsibility for the design and scope of these projects. Because of the close relationships between the sponsoring agencies and the laboratories housing the major animal projects, the projects are highly responsive to the agencies' perceived needs for specific kinds of information.

Communication

Commitments to complete existing long-term studies, shifts in program emphasis, and declining budgets for radiation-related studies that fully occupy staff time and facilities all make it difficult to plan and execute new experiments, to test questions raised by the data as they are being accumulated, or to pursue new goals or new working arrangements, such as closer collaborations with cellular biologists.

Although communication is far from a cure-all for this difficult situation, it can improve the effectiveness of existing projects and the planning of new experiments. A communication system that is functioning well can ensure that research results are widely disseminated. Results from many experiments within many laboratories can be combined to guide planning of the next generation of experiments to ensure that they address the most important scientific questions and produce the greatest amount of useful information.

Recommendations

The raw data that have been generated over the last 35 yr on late radiation effects in animals are the product of an enormous public investment in scientific effort, animals, and money. It has often been recommended that an adequately funded central national archive be established to accommodate this material to make it accessible for continued use. Since many senior investigators in the older projects are approaching retirement age or are being diverted to other work, it is important to act soon to salvage and preserve their materials and to assure safekeeping and future use of the data while the scientists who participated in the studies are available to provide assistance. When new studies are few and must be planned with special care, the existence and availability of such a data bank would be especially advantageous.

Metabolic and radionuclide toxicological data are sifted and organized mainly by the users of such information rather than by those responsible for producing it. Unfortunately, the efforts of such risk-assessment groups appear to be narrowly targeted, hence incomplete and piecemeal. There is a need for thorough compilations and summaries of the vast amounts of data that have accumulated over the past 35 yr in radionuclide research. This undertaking would make the data more accessible and more widely amenable to interpretation.

Environmental Exposures, Radionuclide Transport, and Effects

Historically, some ecologists and other environmental scientists have been concerned with determining the transport of radionuclides in physical media (air, water, soil) and food chains; the potential of certain elements to concentrate (biomagnify) at different ecological levels; and the pathways by which radionuclides may expose individuals or populations, including humans (Reichle *et al.*, 1970).

The committee reviewed all 86 environmental and ecological research projects identified by the Interagency Research Committee. Of these, 71 were sponsored by DOE, 8 by EPA, 3 by DOA, 2 by NRC, and 1 each by NIH and the U.S. Department of Agriculture (USDA).

Objectives of the Research

The major objective of research in progress is the determination of sources, movement, and accumulation of radioactive materials in the environment having an impact on the health and safety of human beings and their environment.

Quality of the Research

Although the scope of environmental research is quite limited, its quality was found to be very high. Principal investigators are well trained and have considerable experience in this type of research. Dissemination of research results to sponsoring agencies via reports and to scientific peers via published literature is very good. Most of this research is important both for regulatory purposes and for expanding scientific knowledge.

Adequacy of the Research

Eighty-five research projects in this field received approximately $9 million of support during fiscal year 1979-1980. They were all involved with the study of radionuclide transport and bioconcentrations. Almost one-third of them deal with the fate of actinides. In addition, one small project is examining effects resulting from exposures to high levels of ionizing radiation.

Thus far, environmental research has provided an adequate basis for setting regulatory standards that ensure acceptable risk under normal conditions. Research is needed on time-dependent radionuclide transport and effects in order to calculate immediate and long-term consequences of accidental radionuclide releases and to establish cost-benefit aspects of remedial action.

Because evaluation of recovery from damage in ecological systems requires months and years of observation, research programs have been terminated too frequently prior to population or ecosystem recovery.

Understanding of recovery processes in damaged ecosystems is incomplete (Cotter and McGinnis, 1965). Because of the failure to conduct studies of recovery following studies of damage, it is likely that ecological effects of ionizing radiations have been exaggerated.

Although there has been considerable descriptive research on responses of plant and animal populations to radiation exposures received under natural conditions (Turner, 1975; Whicker and Fraley, 1974), little analytic research has been conducted on these observations. Results of laboratory studies cannot be expected to explain results obtained under natural field conditions where cumulative stress and synergism affect population behaviors (Miller, 1965). Factorial experiments are necessary to identify additive and synergistic effects involving radiation and other stress factors (Brown and Taylor, 1966; McCormick and McJunkin, 1965).

There is a need to develop nonlinear mathematical and computer models to provide better descriptions of the relations between source factors, dispersal characteristics, biotic processes, and effects. Uncertainties of predictions from models now used by regulatory agencies are due largely to: inadequate field validation of these models; dependence upon experimental results obtained under artifically controlled laboratory studies (National Academy of Sciences, 1981); and efforts to extrapolate site-specific information to regional or generic situations.

Of the 86 ecological and environmental research projects reviewed, only one addresses the effects of external radiation. This single project, with a budget of less than $3,000, is a descriptive study of the recovery of a northern hardwood forest previously exposed to high levels of ionizing radiation. Until additional studies are conducted on the recovery of ecological systems previously exposed to ionizing radiations, accurate estimates of the ecological effects of ionizing radiations are not possible.

There is little research on the additive or synergistic effects of radiation and other environmental stresses on populations and ecosystems. If results of laboratory studies at lower levels of biological complexity are to be extrapolated beyond the laboratory to conditions under which humans and other species actually exist, additive or synergistic studies are an absolute necessity.

Manpower

The most serious inadequacy is the near total absence of manpower training. There is no cadre of young scientists to replace those who were trained between 1950 and 1970. Very shortly, federal agencies with responsibilities for protecting the public from environmental exposure to ionizing radiations will have no source of manpower to replace those lost through attrition. Even now, DOE is unable to hire qualified ecologists to manage ecological or environmental programs.

Recommendations

The committee recommends:

- that longer-term and more broadly focused research projects be developed to improve the quality and applicability of research as well as to contribute to manpower training;
- that research be devoted to the biological and ecological systems affected by ionizing radiation instead of to short-term, narrowly focused research related to a particular radiation source or assessment problem;
- that studies on the uncertainties associated with the use of models of radionuclide transport and dosimetry be given increasing support and that there should be more field validation of model predictions;
- that a modest research program on the ecological effects of ionizing radiations be reinstituted to study additive or synergistic effects of ionizing radiation and other stress factors and recovery of ecosystems previously exposed to high levels of ionizing radiations; and
- that more experimental work be performed on certain dietary pathways of radioactivity in humans, especially on transfers of radioactivity into meat, for which data are especially scarce.

OCCUPATIONAL EXPOSURES

Levels of occupational exposure to radiation are presently well under the limits generally considered to be acceptable (National Council on Radiation Protection and Measurements, 1971). In occupational groups surveyed by the DOE and the NRC, the average dose per worker has remained essentially constant or decreased (U.S. Department of Energy, 1980; U.S. Nuclear Regulatory Commission, 1981). This indicates adherence to present regulations for radiation exposure and suggests that risks from occupational exposures to radiation are quite low.

Nevertheless, the committee believes it would be prudent to continue to reduce all occupational exposures to radiation to the lowest reasonable level. The primary objectives of research concerned with such exposure are listed in the Recommendations portion of this section.

Review of Research

The committee reviewed eight research projects concerning occupational exposures to radiation. Five of them are sponsored by DOE, two by NSF, and one by NRC. These projects are examining calculation for internal dosimetry, hazards from radioactive aerosols, and the development of instrumentation that would be applicable to specific problems.

Objectives

All of the studies reviewed by the committee address themselves in competent fashion to practical problems of occupational dose reduction. The emerging results are being well utilized.

Quality of the Research

The committee believes that the present federal research program to reduce further the possible risks from occupational exposure is of good scientific quality.

Adequacy

The committee believes that the research designed to reduce occupational exposures to radiation, although of high quality, is inadequte in scope. For example, efforts to improve dose measurement or to reduce dose for workers in the medical, mining, and reactor industries are considered minimal.

Recommendations

Practical, cost-effective methods to reduce exposure to radiation in the workplace should be developed.

Three specific areas are particularly deserving of future research:

- Research is needed to improve dosimetry as a measure of the occupational health hazard. Many occupational doses are measured with a single dosimeter and reported as "whole-body dose." To what extent is this appropriate in specific occupations? Research on this subject is necessary to protect workers and to aid epidemiologic studies of the dose-response relationship.
- Research is needed on radon inhalation, especially in the mining industry, and future studies should examine methods to reduce this exposure. In particular, this effort should ensure that controlled laboratory experiments are relevant to the occupational exposure of workers.
- Research is needed on radiation exposures and effects arising from new technologies and from changes in national policies. In particular, improvements are needed in the therapeutic removal of radionuclides from contaminated workers.

Reduction of Radiation Risk From Therapeutic Irradiation

The main justification for continuing research on radiation therapy is its promise of improvements in the treatment of cancer patients. The American Cancer Society estimates that 785,000 new cancers, excluding nonmelanoma skin cancers and carcinoma *in situ*, will be diagnosed in 1980 (Silverberg, 1980). Approximately one-half of these patients will receive radiotherapy at some time, and one-quarter of them—almost 200,000 patients—will have local and/or regional radiotherapy with curative intent as part or all of their initial treatment (Silverberg, 1980).

Significant improvements in local and regional eradication of cancer with radiotherapy were achieved in the 1950's and 1960's with the initiation of high-energy x-rays, high-energy electrons, and advanced treatment-planning methodology. A major result of these advances has been reductions in the risk of normal tissue injury achieved through reduction of normal tissue dose.

Over the past decade, increasing emphasis has been placed on combined modality treatment, i.e., combining radiotherapy with surgery or chemotherapy, or both, to take advantage of the benefits of each modality while simultaneously attempting to decrease the adverse effects of maximum treatment by each modality. Despite modest improvements gained with combined modality therapy, there is a strong need for increased support of research in radiotherapy, along with the associated radiation biology and radiological physics, because tumors recur locally or regionally in approximately 100,000 patients each year (Pistenma, 1980). Significant improvement may result from one or more of the areas of investigation in radiotherapy, either alone or as part of a well-designed combined modality treatment regimen.

Only a portion of the radiation administered to cancer patients is effectively concentrated on the tumor. Healthy tissues of the patient are also irradiated during the treatment, thereby subjecting them to potential damage. Consequently, it was a main concern of the committee to review research aimed at improving the ratio of the tumor radiation dose to total patient dose. The findings of the committee's review of 27 research projects in this area are discussed below. Almost all the research conducted in this area and all the research reviewed by the committee is sponsored by the National Cancer Institute (NCI).

Review of Research

For a given tumor control probability, a main goal of radiation therapy is to reduce the dose to normal tissue to levels as low as practicable in order to minimize morbidity and other risks from exposure to radiation. A measure of the successful achievement of these goals is the degree to which the ratio of tumor dose to the dose received by healthy tissue is increased, thereby decreasing risk of various undesirable radiation side-effects, including radiation oncogenesis. This can be achieved by various single or combined approaches involving physical, chemical, biological, and other factors.

The committee's study led to some recommendations concerning the review process that results in the awarding of grants and contracts by government agencies, the goals of future research, and practices of funding. As the use of radiation therapy increases and advances in the future, leading to further improvement in tumor cure rates and general cancer control, more attention should be paid to the reduction of unavoidable irradiation of normal tissues so that latent radiation risk to patients cured of their disease can be reduced. The committee noted that the radiation therapy community, through a series of extensive studies from 1975 to 1980, defined a number of research objectives (Pistenma, 1980). These objectives, with some additions, appear to have been exceedingly well drawn and are strongly endorsed by the committee.

The 27 projects reviewed by the committee are involved with such topics as heavy particle therapy; conventional radiation therapy; instrumentation developments for heavy particle, conventional megavoltage, and high-energy radiation therapy; intraoperative therapy using high-energy electrons; slow neutron-boron capture radiation therapy; dosimetry in various aspects of radiation therapy; and related radiobiological topics.

The heavy particle therapy projects included the use of negative pi-mesons, heavy ions, and protons. Projects on fast neutron therapy were also reviewed. The instrumentation projects included studies of three-dimensional tomography with ultrasound, development of tissue localization and treatment planning techniques for heavy particle therapy, development of a DT (deuterium, tritium) neutron generator for radiation therapy, and development of computed tomographic (CT) scanner for simulation and radiation therapy. Several projects concerned different aspects of fundamental radiobiology. Other projects were concerned with instrumentation and developments for conventional high-energy x-ray and electron-beam therapy, with attention to different aspects of treatment-planning and corrections for inhomogeneities. The dosimetry projects included studies on neutron dosimetry, a combination of chemical and colorimetric dosimetry, basic neutron absorption parameters that are necessary for fast neutron therapy, and physical characteristics of heavy ions for heavy particle therapy.

Objectives

As previously stated, a major objective of radiation therapy is to improve tumor control rates while simultaneously minimizing the radiation risk to the patient by reducing as much as practical the unavoidable irradiation of normal tissues. Such dose reduction increases the tolerable radiation doses to the patients and reduces the probability of latent radiation-related effects developing in patients cured of their initial disease. Associated objectives are the minimization of occupational exposures of personnel administering radiation treatment as well as exposures of the general public who visit treatment facilities, both of which are expected to follow a reduction in patient doses.

Quality of the Research

The research projects reviewed were generally found to be of good-to-excellent quality, with several exceptions. Analysis of the exceptions indicated that if the government agencies involved in such research had recognized the need for multidisciplinary reviewers in a field as complex as radiation therapy and its associated areas, award actions might have been different. The committee's challenges to some of the projects in this sample provide an indication of the fallibility of the peer review process and reaffirm earlier observations that improvements in that process may well be in order (National Academy of Sciences, 1980a). Its recommendations for improved use of research funds appear below.

Adequacy

The attainment of the objectives of radiation therapy necessitates a multidisciplinary approach utilizing physical, chemical, biological, and other factors. Most of the projects reviewed were generally of good quality. The exceptions mentioned above represented only a fraction of current research in this area and do not touch on all of the topics of interest. In addition to research investigating new techniques of tumor irradiation, studies in radiation therapy need to be expanded in the areas of combined modality of treatment, hyperthermia, radiosensitizers, radioprotectors, radiation immunology, and radiation toxicology.

Recommendations

- The committee supports the research recommendations developed by several study groups within the radiation therapy community from 1975 to 1980 (Committee for Radiation Oncology Studies, 1976, 1978; Committee on Radiation Oncology Studies, 1978; U.S. Department of Health, Education, and Welfare, 1976; Rubin *et al.*, 1979).
- The committee recommends that radiation research conducted

by government laboratories be subject to the same intensity of review as are outside laboratories supported by government funds. Intramural research funded by the government and conducted in its own laboratories, as well as interagency transfer of funds to support research, would benefit from the same type of peer review used for extramural research.

• In the interests of long-range productivity and the attraction and retention of competent scientists in radiation-related research, the committee recommends that attention be given to the concept of stability and continuity in research programs. This might be accomplished by more frequent awards of 5-yr contracts for both new and renewal projects that peer groups have judged to have a high probability of producing significant results. Longer funding cycles have the additional advantage of reducing the frequency of peer reviews, which would result in time-saving benefits for all concerned. The committee recognizes that this problem is pervasive in other fields of research, but has confined its recommendations to the subject of this report.

• Radiation-induced carcinogenesis in radiation therapy patients should be evaluated, and recommendations for modifying such treatment to reduce the incidence of such carcinogenesis should be made. The evaluative studies should be both retrospective and prospective and should include not only factors in radiation treatment but also the influence of chemotherapy in combined programs with respect to initial appearance of cancer.

Reduction of Radiation Risk From Diagnostic Radiation

The diagnostic use of radiation in medical practice constitutes the largest source of ionizing radiation to which humans are exposed in the United States (National Academy of Sciences, 1980b). An estimated 105 million persons receive one or more x-ray examinations annually. This use contributes an estimated 17 million person-rems annually, which is 45% of the total population dose of approximately 42 million person-rems. In nuclear medicine, radiopharmaceuticals are given to 10 to 12 million patients annually, contributing 3 million person-rems to the total population dose.

The committee is interested primarily in research directed toward the reduction of patient dose in diagnostic radiology and improvement in the quality of diagnostic information. Such technological improvements have been achieved or are being pursued in the following research areas:

• the development of new or improved radiation sources and detection equipment and image-enhancing techniques;
• utilization of low-attenuation materials and better scatter rejection techniques; or
• use of nonionizing radiation techniques such as ultrasonics or nuclear magnetic resonance techniques to replace x-rays.

In recent years, substantial reductions in radiation exposure and advances in diagnostic capability have been observed. Further advances seem attainable.

Review of Research

The committee reviewed federally supported research aimed at reduction of risk from diagnostic radiation. The principal agencies responsible for programs in this area include the NIH, the Food and Drug Administration (FDA), DOE, the National Science Foundation (NSF), and the VA.

The committee reviewed 60 projects listed by the various agencies. They all involved medical diagnostic applications of ionizing radiation. Approximately half of these projects were sponsored by NIH, while the remainder were more or less equally distributed among DOE, FDA, VA, and NSF. Various BRH intramural projects in medical radiation and nuclear medicine were also reviewed. The committee believes that it reviewed a sufficient number of projects within a wide enough range to constitute a representative sample of the federal effort in the area of diagnostic radiology, nuclear medicine, and associated instrumentation physics and engineering.

Objectives

The federally supported research projects on medical diagnosis with ionizing radiation were reviewed in the perspective of the broad research objectives indicated in the following important questions:

• What is the impact of the newer technologies in radiology and nuclear medicine with regard to improved diagnostic capability and reduced radiation exposure, and how is it assessed?
• How can the trade-offs between patient dose, information content of the x-ray image, and cost-effectiveness of the imaging process be better defined and evaluated?
• What research is required to examine effectively the extent to

which imaging processes using ionizing radiation can or should be replaced or supplemented by imaging processes using nonionizing modalities such as ultrasonography or nuclear magnetic resonance?

- What research is needed to improve the efficiency with which photons are utilized in medical imaging processes?
- What levels of manpower and training in research are required to promote greater diagnostic information and reduced radiation exposure in the fields of radiology and nuclear medicine?

Quality of the Research

As a result of federally and industrially supported research, the radiological imaging field is experiencing a phenomenal growth in new sophisticated electronic and computer instrumentation. This is resulting in a significant reduction in radiation exposure and in improved diagnostic capability.

The overall quality of research in the radiological sciences, based on reviews by committee members, consultants, and other scientists, was found to be good but reflected some unevenness.

The most promising and exciting new areas of federally supported research in the radiological sciences encompass at least the following: computed tomography, including dynamic and quantitative aspects of transmission and emission; photoelectric imaging, including digital image analysis; magnification radiography; and alternative imaging modalities, including ultrasonography and nuclear magnetic resonance.

The major support for several of the advanced technologies (especially computed tomography) has emanated from industrial and private sources. This focus of funding has resulted in the marketing and promulgation of the technologies without a meaningful, coordinated assessment of their capacity for dose reduction, improved diagnosis, or cost-effectiveness. If the government is to exercise a role in controlling the growth and utilization of this technology and regulating the accompanying exposures to radiation, it must assume a more central role in supporting its development and the assessment of its capabilities, effectiveness, and future applications.

In other areas of medical instrumentation, the committee has identified further opportunities for improvements in diagnosis at equal or lower radiation exposure, for example advanced detection technology and improved scatter rejection techniques. Without federally supported research programs promoting these developments or demonstrating their utility, however, industrial and private funding bodies may be insufficiently motivated to invest the necessary resources to bring these technologies into the clinical realm.

Adequacy

Important research not receiving sufficient attention and support includes the development of advanced imaging instrumentation to reduce radiation and improve diagnoses and the evaluation of the efficacy of new ionizing radiation and alternative nonionizing radiation diagnostic procedures and instrumentation as it affects diagnostic capability, dose reduction, or cost benefit.

Recommendations

• To ensure that the best and most cost-effective radiological technologies for dose reduction and improved diagnoses are made available to the public, the committee recommends that adequate federal support for research on diagnostic imaging be maintained. More adequate funding, commensurate with the increase in expenditures for radiological equipment and procedures within the public and private sectors, will improve the quality and productivity of research by adding more stability to successful teams and by raising the probability of funding excellent research proposals.

• Investigator-initiated research grants and agency-directed contracts are both approaches to research management having unique and desirable qualities that should be retained. Although a balance should be maintained, the committee recommends that a greater fraction of total support should be mediated through the granting mechanism, primarily because of the broader scope of research ideas and the more demanding evaluation provided by peer review.

• Adequate federal support for radiological diagnostic research training is recommended to ensure the proper utilization of diagnostic imaging procedures and to minimize radiation risks to patients.

• A focus for management and coordination of research programs and for allocation of funds in the radiological sciences should be established to ensure a high-quality and balanced research effort.

References

Brown, G. N., and F. S. Taylor, Jr. 1966. Interaction of radiation and photoperiodism in *Xanthium chinense:* Effects of variation in time of radiation with respect to photoinduction. Radiat. Bot. 6:145-150. Committee for Radiation Oncology Studies. 1976. Research plan for radiation oncology. Cancer 37(4 Suppl.):2031-2148.

Committee for Radiation Oncology Studies. 1978. Proposal for a program in particle-beam radiation therapy in the United States—A report from the Committee for Radiation Oncology Studies (CROS) and its Particle Subcommittee. Cancer Clin. Trials 1:153-208.

Committee on Radiation Oncology Studies. 1978. Computed tomography in radiation therapy: Report of the Committee on Radiation Oncology Studies, Subcommittee on CT Scanning and Radiation Therapy. Int. J. Radiat. Oncology Biol. Phys. 4:313-324.

Cotter, D. J., and J. T. McGinnis. 1965. Recovery of hardwood stands 3-5 years following acute irradiation. Pp. 1663-1674 in F. P. Hungate, ed. Radiation and Terrestrial Ecosystems. Pergamon Press, New York.

Harley, J. H. 1980. Measurements of Environmental Radioactivity. Paper presented at a FREIR Committee workshop held in Albuquerque, N. Mex., April 14-16, 1980. National Research Council, Washington, D.C. 18 pp.

McCormick, J. F., and R. E. McJunkin. 1965. Interactions of gamma radiation and other environmental stresses upon pine seeds and seedlings. Health Phys. 11:1643-1652.

Miller, L. N. 1965. Changes in radiosensitivity of pine seedlings subjected to water stress during chronic gamma irradiation. Health Phys. 11:1653-1662.

National Academy of Sciences. 1980a. Review of Energy Research on Human Health Effects of Low Doses of Ionizing Radiation. Report of the Committee on DOE Research on Health Effects of Ionizing Radiation. National Academy of Sciences, Washington, D.C. 93 pp.

National Academy of Sciences. 1980b. The Effects on Populations of Exposure to Low Levels of Ionizing Radiation. Report of the Committee on the Biological Effects of Ionizing Radiations. National Academy of Sciences, Washington, D.C. 638 pp.

National Academy of Sciences. 1981. Testing for Effects of Chemicals on Ecosystems. Committee to Review Methods for Ecotoxicology, Commission on Natural Resources. National Academy Press, Washington, D.C. 98 pp.

National Council on Radiation Protection and Measurements. 1971. Basic Radiation Protection Criteria. Recommendations. NCRP Report No. 39. National Council on Radiation Protection and Measurements. 135 pp.

National Council on Radiation Protection and Measurements. 1978a. A Handbook of Radioactivity Measurements Procedures with Nuclear Data for Some Biologically Important Radionuclides. Recommendations. NCRP Report No. 58. National Council on Radiation Protection and Measurements, Washington, D.C. 506 pp.

National Council on Radiation Protection and Measurements. 1978b. Instrumentation and Monitoring Methods for Radiation Protection. Recommendations. NCRP Report No. 57. National Council on Radiation Protection and Measurements, Washington, D.C. 177 pp.

Pistenma, D. A. 1980. Radiotherapy. Pp. 91-115 in Science Projection Papers Working Documents. March 11, 1980. Vol. II. A proposed Federal Radiation Research Agenda. Public Meeting. U.S. Department of Health, Education, and Welfare, Public Health Service, National Institutes of Health, Bethesda, Md.

Reichle, D. E., P. B. Dunaway, and D. J. Nelson. 1970. Turnover and concentration of radionuclides in food chains. Nuclear Safety 11(1):43-55.

Rubin, P., R. B. Cowen, and D. J. Rubin, eds. 1979. The radiation oncology research program: Recommended research proposals. Int. J. Radiat. Oncology Biol. Phys. 5:593-773.

Silverberg, E. 1980. Cancer statistics, 1980. CA—Cancer J. Clinicians 30(1):23-38.

Turner, F. B. 1975. Effects of continuous irradiation on animal populations. Radiat. Biol. 5:83-144.

U.S. Department of Energy. 1980. Eleventh Annual Report of Radiation Exposures for DOE and DOE Contractor Employees. DOE/EV Report 0072/11. National Technical Information Service, Springfield, Va.

U.S. Department of Health, Education, and Welfare. 1976. National Cancer Program. Digest of Scientific Recommendations. Radiation Oncology. NCP Publ. No. TD-P1-76. Public Health Service, National Institutes of Health, National Cancer Institute, Bethesda, Md. 81 pp.

U.S. Nuclear Regulatory Commission. 1981. Eleventh Annual Occupational Radiation Exposure Report-1978. NUREG Report 0593. National Technical Information Service, Springfield, Va.

Whicker, F. W., and L. Fraley, Jr. 1974. Effects of ionizing radiation on terrestrial plant communities. Adv. Radiat. Biol. 4:317-366.

10
Research Management

This chapter examines the management of the federal government's research programs on the effects of ionizing radiation. Information was gathered by three methods. (1) The committee surveyed federal agencies to determine the scope of their current research activities on the biological effects of ionizing radiation. (2) Members of the committee and consultants conducted interviews with past and current agency officials, congressional staff, union officials, and representatives of groups concerned with ionizing radiation matters to determine their views on the management problems affecting the research under study. (3) The reviewers of the scientific programs reported on management problems discovered during the review process.

History of Federal Government Involvement in Ionizing Radiation Research

Scientific research in the field of ionizing radiation was pursued largely in the private sector following the discoveries of roentgen rays by Wilhelm Conrad Roentgen in 1895 and the advent of World War II. Government involvement in such research was indirect, being limited essentially to the support of educational, research, and health care institutions that might have had an interest in research on ionizing radiation.

Following the establishment of the Manhattan Project and the development of the atomic bomb by the United States, the federal government became increasingly involved in the development, application, regulation, and control of ionizing radiation. A number of institutional arrangements evolved among federal agencies, private industry, and educational and research institutions in order to conduct radiation research in support of government missions. Direct participation of the government was maintained at different levels, depending upon the specifics of the arrangements.

Prior to the federal government's initial investment in the development of the atomic bomb, major interest in ionizing radiation was centered in medical, academic, and industrial sectors where much of the research and development in ionizing radiation was pursued. Since there was early evidence of radiation hazards, such as acute skin burns and, later, skin tumors, scientific interest in radiological protection was stimulated. This led to the establishment, in 1928, of an International Commission on Radiological Protection (ICRP), followed the same year by the creation of an American counterpart organization, the Advisory Committee on X-Ray and Radium Protection, which became chartered by the Congress of the United States as the National Council on Radiation Protection and Measurements (NCRP) in 1964. Both organizations were, and are, composed of groups of scientists in the private sector who study and report on various aspects of protection against ionizing radiation. The NCRP has played a major role in the analysis of data and dissemination of information in the field of radiation protection since its inception and remains a singularly active and productive nongovernmental institution today.

In December 1938, the discovery of nuclear fission launched the new field of atomic energy. By 1939, Leo Szilard, Enrico Fermi, and their colleagues had raised the possibility of creating a controlled explosive devise using atomic energy. By 1941, the National Academy of Sciences' Uranium Committee was actively examining the subject.

Between 1943 and 1947, the federal government invested large sums of money in the development of nuclear weapons. A portion of these funds was supplied directly by the President's emergency funds, part of which was authorized to be expended for confidential projects. These funds were dispensed through the National Defense Research Council (NDRC) and the Office of Scientific Research and Development (OSRD). The larger portion of the cost of this research and development program was borne by the budget of the War Department for the Manhattan Engineering District during fiscal years 1943 through 1947. The Military Appropriation Act for fiscal year

1947 bears the first public reference to atomic energy in this series of appropriations.

Through the wartime research and development efforts, not only was a large base of physical and biomedical information created, but also a complex set of relationships was established among the federal government agencies, the National Academy of Sciences, the academic scientific community, and industry. These relationships persist in many forms to this day.

The first military use of an atomic weapon occurred in 1945. In recognition of the potential of the technology and with the military research experience as a guideline, the Congress enacted the Atomic Energy Act of 1946 (U.S. Public Law 585, 1946). Section 1(b) states that the purpose of the Act is to establish the following programs:

> (1) A program of assisting and fostering private research and development to encourage maximum scientific progress;
>
> (2) A program for the control of scientific and technical information which will permit the dissemination of such information to encourage scientific progress, and for the sharing on a reciprocal basis of information concerning the practical industrial application of atomic energy as soon as effective and enforceable safeguards against its use for destructive purposes can be devised;
>
> (3) A program of federally conducted research and development to assure the Government of adequate scientific and technical accomplishment;
>
> (4) A program for Government control of the production, ownership, and use of fissionable material to assure the common defense and security and to insure the broadest possible exploitation of the fields; and
>
> (5) A program of administration which will be consistent with the foregoing policies and with international arrangements made by the United States, and which will enable the Congress to be currently informed so as to take further legislative action as may hereafter be appropriate.

To achieve these goals, the Act created the Atomic Energy Commission (AEC) and empowered it with the responsibility of conducting research and development programs directed at such subjects as the use of fissionable and radioactive materials for medical, biological, or military purposes and the protection of health during research and production of fissionable materials [Section 3(a)]. The AEC was also charged with establishing and operating its own facilities to produce fissionable material and was given the responsibility of acquiring and maintaining all fissionable material in the United States. Moreover, the newly created agency was empowered to authorize the possession of, but not the title to, fissionable material by others and to distribute fissionable material for research, development, or medical therapy to organizations holding an AEC license to do so. The Act specified that the agency should conduct research and development in the military application of atomic energy. Furthermore, it stipulated that the AEC require licensure for the use of fissionable material or atomic energy and for the manufacture, production, or export of any equipment or device using such materials.

A further refinement of these concepts was embodied in the Atomic Energy Act of 1954 (U.S. Public Law 83-703, 1954), which required the AEC to conduct research and development for both military and peaceful uses of atomic energy. This Act specified four categories of research: military applications; processes, materials, and devices that can be used to produce nuclear energy to generate electricity; safety during research and production of fissionable materials; and medical, biological, agricultural, or health purposes.

It was through this series of legislative developments that both major governmental roles in the atomic field were defined: research for production and research for protection. They were combined in the charge to a single agency, the Atomic Energy Commission.

Although the U.S. Armed Forces occupying Japan, using both Japanese and American scientists, gathered initial data to determine acute medical effects in the radiation-exposed population of Hiroshima and Nagasaki, it soon became evident that a long-term integrated study would be necessary. In 1947, the Atomic Bomb Casualty Commission (ABCC) was formed within the National Academy of Sciences (NAS). First supported by the U.S. military occupational forces, this commission beginning in 1949 received funds from the AEC. This activity has continued to date under the auspices of successor agencies to the AEC, now the Department of Energy (DOE). On April 1, 1975, the program was transformed from a project directed and funded entirely by the United States to a binational study under joint Japanese and American direction and funding. It was renamed the Radiation Effects Research Foundation (RERF). The U.S. effort continues within the NAS.

Independent of governmental initiatives, and in response to nuclear weapons testing and public concern about the potential effects of ionizing radiation on human populations, the NAS, with support from the Rockefeller Foundation, formed the Committee on Biological Effects of Atomic Radiation (BEAR) to study this subject. The BEAR Committee issued a series of reports between 1956 and 1963.

As evidence of still broader concern, the General Assembly of the United Nations in 1955 established the United Nations Scientific Committee on the Effects of Atomic Radiation (UNSCEAR). Among other tasks associated with monitoring and assembling reports of radiation exposure throughout the world, UNSCEAR was required "to make yearly reports and to develop . . . a summary of reports received on radiation levels and radiation effects on man and his environment" (United Nations, 1969). Periodic reports by this group have provided reviews of the worldwide scientific information and opinion concerning the exposure of humans to atomic radiation.

In 1959, the federal government addressed its own concerns about the effects of radiation on health by creating the Federal Radiation Council (FRC) within the Executive Office of the President to provide a federal policy on radiation exposure of humans. A major function of the FRC was to "advise the President with respect to radiation matters, directly or indirectly affecting health, including guidance for all Federal agencies in the formulation of radiation standards and in the establishment and execution of programs of cooperation with States. . . ." (U.S. Public Law 86-373, 1959).

The FRC consisted of the Secretaries of the Departments of Health, Education, and Welfare; Defense; Commerce; Labor; and the Chairman of the AEC (or their designees) and others appointed by the President of the United States. It was charged with consulting scientists and experts in radiation matters, including the President of the NAS, the Chairman of the NCRP, and qualified experts in biology, medicine, and health physics.

The same bill that created the FRC also provided that the AEC could enter into agreements with the state governments authorizing them to regulate radioactive materials, i.e., "(1) by-product materials; (2) source materials; (3) special nuclear materials in quantities not sufficient to form a critical mass," for the protection of the public health and safety from radiation hazards (U.S. Public Law 86-373, 1959). This provision curtailed some of the regulatory authority of the AEC.

In February 1970, the FRC asked the NAS to undertake a complete review and evaluation of existing scientific knowledge concerning radiation exposure of human populations. The product of that study was the 1972 report of the Committee on the Biological Effects of Ionizing Radiation (often called the BEIR I Report) (National Academy of Sciences, 1972). This committee examined all evidence pertaining to the effects of ionizing radiation on human populations and provided conservative estimates of the risks of untoward effects of low levels of radiation exposure. Subsequently, these estimates were used by regulatory agencies in setting standards for limiting occupational and public exposure.

The Environmental Protection Agency (EPA) was established by the President's Reorganization Plan No. 3, which was described in President Nixon's message to Congress on July 9, 1970 (U.S. President, 1970). Among the various functions assigned to the EPA were those of the FRC and the environmental standard functions of the AEC. The President discussed EPA's broad mandate to control environmental pollution and to enhance the environment and described specific activities of the FRC and AEC:

> Environmental radiation standards programs—The Atomic Energy Commission is now responsible for establishing environmental radiation standards and emission limits for radioactivity. Those standards have been based largely on broad guidelines recommended by the Federal Radiation Council. The Atomic Energy Commission's authority to set standards for the protection of the general environment from radioactive material would be transferred to the Environmental Protection Agency. The functions of the Federal Radiation Council would also be transferred. AEC would retain responsibility for the implementation and enforcement of radiation standards through its licensing authority.

This administrative move separated, for the first time, responsibilities for protection from those for development of applications. On December 2, 1970, the transfer of functions was formally accomplished and the EPA's Office of Radiation Programs was created.

Under the terms of the new method of administration, the establishment of standards for the protection of the general environment from radioactive materials was an EPA function. Enforcement fell within the purview of the AEC licensure process. Based upon a variety of subsequent authorities, the EPA undertook the regulation of the discharge of radioactive material into navigable waters (Federal Water Pollution Control Act; U.S. Public Law 86-70, 1959), to protect drinking water supplies when the states failed to do so (Safe Drinking Water Act; U.S. Public Law 93-523, 1974), to regulate the recovery and disposal of all radioactive wastes (Resource Conservation and Recovery Act; U.S Public Law 94-580, 1976), and to regulate airborne emissions of radioactive materials (Clean Air Act; U.S. Public Law 88-206, 1963).

As further changes in administration of radiation programs evolved, the controls were applied directly by the Department of Defense (DOD) and the DOE in their own operations.

The Energy Reorganization Act of 1974 (U.S Public Law 93-438, 1974) dissolved the AEC entirely and created two new agencies to assume the AEC's remaining functions: the Nuclear Regulatory Commission (NRC), which assumed the AEC licensing and remaining regulatory functions, and the Energy Research and Development Administration (ERDA), which was charged with conducting research and development for both military and peaceful uses of atomic energy. When the responsibility for setting standards became the responsibility of the EPA and the licensing and regulatory activities were assigned to the NRC, the control functions pertaining to radiation became further separated from the activities related to research, development, and application of atomic energy. Moreover, the legislation permitted some regulatory powers of the NRC to be relinquished to a state if that state's programs were adequate and comparable to those of the NRC.

The Department of Energy Organization Act of 1977 (U.S. Public Law 95-91, 1977) created the Department of Energy (DOE) and transferred to it the responsibilities of ERDA. Under Title III, Sec. 301(a) of this Act, the DOE was given the responsibility for developing and producing nuclear weapons systems for the DOD in facilities owned by DOE, but operated by private companies and universities, e.g., the Los Alamos Scientific Laboratory in New Mexico and the Lawrence Livermore Laboratory in California, both operated by the University of California; Rocky Flats in Colorado, operated by Rockwell International; and Pantex in Texas, operated by Mason and Hanger. Furthermore, it was charged with developing peaceful applications of nuclear energy and technology, especially power sources, in cooperation with private companies, and radioisotopes for medical and industrial applications, primarily within the national laboratories originally established by the AEC.

In addition to the above chains of responsibility, several other federal organizational functions were and are active in parallel with the AEC-ERDA-DOE and FRC-EPA activities:

- The Naval Nuclear Propulsion Program, jointly operated by the DOD and DOE, develops and supports the reactors for a fleet of nuclear-powered submarines and surface ships.
- Health care facilities dealing with ionizing radiation are administered by the DOD, the Veterans Administration (VA), and the Department of Health and Human Services (DHHS), formerly the Department of Health, Education, and Welfare (DHEW). The DHHS sponsors research involving the use of radiation technology in both health care and basic radiation biology.
- The Food and Drug Administration (FDA) has the responsibility to provide guidance concerning the use of food and animal feeds

containing radionuclides. It also regulates radiopharmaceuticals and radiation-related medical devices, and sets performance standards for diagnostic x-ray machines and other electronic products that emit radiation.

• Control of consumer products that are a source of ionizing radiation exposure is shared by the FDA, the NRC, and the Consumer Product Safety Commission (CPSC), depending on the specific product.

• The National Aeronautics and Space Administration (NASA) is responsible for the development of nuclear aerospace applications.

• The transport of radioactive materials or goods is coordinated by the Materials Transportation Bureau, using authorities of the NRC, the Department of Transportation (DOT), and the Postal Service.

• As a separate function, the following federal agencies are concerned with monitoring and regulating occupational exposures of specific groups to radiation: NRC, workers of licensees; the Department of Labor's Mine, Safety, and Health Administration (MSHA), miners exposed to radioactive materials; and the Occupational Safety and Health Administration (OSHA), standards for workers other than those covered by NRC and MSHA. OSHA may delegate authority to states that meet OSHA criteria. The National Institute for Occupational Safety and Health (NIOSH) conducts research in support of OSHA regulatory activities.

• Apart from the health care and radiobiological research functions of the DHHS, the department conducts epidemiologic studies under the aegis of the NCI, FDA, NIOSH, and CDC. Data collection and analysis is conducted by the National Center for Health Statistics.

• The National Bureau of Standards has a responsibility for establishing and maintaining reference bases for measurements, data, and materials and for providing infrastructure services for the physical and engineering sciences.

• The Department of Defense deals with radiation research relevant to its primary military mission; hence, most of its research is directed toward the effects of high levels of radiation exposure and those effects principally occurring in materials and in biological systems.

• The U.S. Department of Agriculture uses ionizing radiation as a research tool for the development of new plant strains, especially food grains, and the production of sterile male insects used in pest control systems.

• The National Science Foundation is charged with the support of research in basic science generally. Its ionizing radiation research component tends to not involve human health effects studies, but, rather, is directed toward fundamental biological issues in which radiation is generally a tool used in specific biological investigations.

Despite the appearance of an orderly distribution of separate functions—developmental, regulatory, and basic research—overlapping interests have resulted in a less clearcut discrimination of activities among the federal agencies. The most recent attempt to coordinate these efforts was made by the Interagency Task Force on Ionizing Radiation, which was convened by the Secretary of DHEW in June 1978 at the request of the Executive Office of the President. To some degree, each function can be described as pertinent to more than one organization. More importantly, for the purpose of this report, the *research* required for development and protection is not always supported by the agency that will have principal need for the information. Furthermore, it is not clear whether basic research in ionizing radiation can be categorized as either developmental or protectionist so that the projects can be assigned to one or another agency whose missions involve the application of such research. The problem of the moment is also an opportunity: rational decisions must be made to determine the research to be conducted and applied in terms of institutional responsibilities.

Current Federal Government Management of Radiation Research

The large number of federal executive agency interests in radiation research are reflected in the membership of the Interagency Radiation Research Council (IRRC), which was created by President Carter on February 21, 1980 (U.S. President, 1980) at the recommendation of the Interagency Task Force on the Health Effects of Ionizing Radiation (Department of Health, Education, and Welfare, 1979). The IRRC is composed of representatives from all 14 federal agencies having significant research, operational, and protective functions in the area of radiation. These agencies are DOE, DHHS, DOD, DOT, NRC, CPSC, EPA, NASA, the Department of Agriculture, the Department of Commerce (DOC), the Department of Labor (DOL), the Federal Emergency Management Agency (FEMA), the National Science Foundation (NSF), and the Veterans Administration (VA).

Also established in 1980 by executive order is the Radiation Policy Council (RPC), which is charged with formulating and implementing federal policy relating to radiation protection. Its membership includes representatives of the DOD, DOC, DOL, DHHS, DOT, DOE, VA, EPA, NSF, FEMA, NRC, Housing and Urban Development (HUD), and Department of Justice (DOJ).

Agencies with Authority to Conduct Research but that Reported No Activity

Responses from all federal agencies to a letter of inquiry from the Director of the National Institutes of Health concerning current activity in support of research on the effects of ionizing radiation included those from several agencies that were not currently engaged in such work. The agencies, the respondents, and any significant comments from the agencies follow:

Central Intelligence Agency, James H. McDonald, Director of Logistics: ". . . not currently engaged in any research efforts concerning ionizing radiation or its related biological effects."

Consumer Product Safety Commission, Susan B. King, Chairman: "Our involvement with the regulation of products containing ionizing radiation sources has been minimal due to our limited jurisdiction in this area. . . ."

U.S. Department of Housing and Urban Development, Donna E. Shalala, Assistant Secretary for Policy Development and Research: "We will rely on your Department [DHEW] and the Environmental Protection Agency to provide regulations to be used to protect populations who may be at risk because of specific environmental exposure to sources of low-level ionizing radiation."

U.S. Department of the Interior, Joan M. Davenport, Assistant Secretary of the Interior: ". . . no research of this nature being conducted either in-house or by contract."

U.S. Department of Labor, Robert Copeland, Director, Office of Health and Disability: "We have programs which are designed either to limit exposure of workers to ionizing radiation or to compensate those workers who may have become ill or died as a result of exposures. Research with regard to the former area is primarily performed for us by NIOSH, though it is possible that some kinds of research may need to be performed by DOL in the future in support of any new standard proposed. For example, OSHA or MSHA might need to perform an economic impact analysis which would deal with questions such as the numbers of workers exposed and the technological feasibility of compliance.

"With regard to compensation, the Employment Standards Administration (ESA) currently has a contract to design, if possible, medical and disability standards that could be used in making determinations of compensability under the Federal Employees' Compensation Act."

The comments have been included to illustrate the breadth of interests in the research efforts, some of their potential applications within the federal government, and the interdependence of agencies and departments upon one another, as well as the selectivity of resource sources.

CONGRESSIONAL COMMITTEES

In parallel with the multiplicity of interests in the Executive Branch of government, many necessarily overlapping both the fields of supported research and the research information required for application, similar collections of legislative interests are represented among the committees of the Congress of the United States. These are listed in Tables 10-1 and 10-2 with examples of the subjects that were studied by each committee.

Absent from these tables are the appropriations committees of Congress through which authorizations become functioning realities by the process of funding and, similarly, authorization may be given lower priority by funding limitations. The specific agency oversight committees are also excluded from these tables since their interests clearly correspond with those of the executive agencies.

ORGANIZATION OF THE FEDERAL RESEARCH EFFORT

As discussed above, the recent history of federal involvement in ionizing radiation research and regulation activities is characterized by an increasing dispersal of managerial authority. As late as the mid-1950's, the AEC was clearly the dominant agency in the field, holding sole responsibility for nearly every task relating to the control of the hazards of ionizing radiation. Today, there are literally dozens of agencies that can legitimately claim a voice in the formulation of policy pertaining to one or more aspects of ionizing radiation. Where once there was an extraordinary degree of centralization in managerial authority, there is now an extreme degree of jurisdictional fragmentation. It is doubtful that any other program area in government has been subjected to such a marked change in jurisdictional alignment in as short a time.

TABLE 10-1 Senate Committees Concerned with Radiation

Committees	Subject of Legislation
Labor and Human Resources Committee, Subcommittee on Health and Scientific Research	Radiation exposure
	U.S. liability resulting from Nevada test site activities (referred jointly to Labor and Human Resources Committee and Judiciary Committee)
	Medical and dental exposure to radiation
Governmental Affairs Committee, Subcommittee on Energy, Nuclear Proliferation, and Federal Services	Coordination of federal agencies involved in radiation protection and radiation research
Environment and Public Works Committee	Nuclear power plant safety
Energy and Natural Resources Committee, Energy Regulation Subcommittee	Nuclear Waste Policy Act (U.S. Congress, 1979)

This dispersal of authority is the result of a number of distinct causes. The dominance of the AEC in nuclear affairs was successfully challenged by those who believed that the promotion and regulation of technological activities should be vested in separate agencies. The Joint Committee on Atomic Energy, which had centralized congressional oversight of radiation matters, was disbanded in 1977, giving other congressional committees the opportunity to initiate and monitor legislation affecting agencies concerned with radiation. Continued demonstration of the efficacy of radiation in medical diagnosis and therapy and its importance in basic biological investigations led to a diffusion of radiation-related research among the various federal health research agencies. And a growing concern with environmental and occupational hazards and risks to the consumer brought about the establishment of new federal agencies, some of which were assigned jurisdiction for monitoring and regulating radiation sources.

As the number of agencies involved in radiation-related activities grew, so did a belief that there was a need to coordinate agency policies in this field. It was feared that agencies would promulgate

TABLE 10-2 House Committees Concerned with Radiation

Committees	Subject of Legislation
Interstate and Foreign Commerce Committee, Subcommittee on Health and the Environment, and Subcommittee on Energy and Power	Radiation exposure
	Nuclear power plant safety (Joint referral to Interior and Insular Affairs Committee and Interstate and Foreign Commerce Committee)
Interior and Insular Affairs	Nuclear power plant safety (Joint referral to Interior and Insular Affairs Committee and House Interstate and Foreign Commerce Committee)
	Nuclear Regulatory Commission, risk analysis
	Nuclear safety research
	Radium pollution in Colorado
Science and Technology Committee, Energy Research and Production Subcommittee	Nuclear Regulatory Commission, safety research (Joint referral to Science and Technology Committee and Interior and Insular Affairs Committee)
	Nuclear reactor safety (DOE)
	Nuclear waste research
Judiciary Committee	Hiroshima and Nagasaki explosion, medical expenses
Armed Services Committee	Nuclear energy emergencies: evacuation plans, explosion-proof buildings
Government Operations Committee	Radiation: EPA responsibility

inconsistent radiation control standards, issue contradictory statements on radiation health effects, and sponsor needlessly duplicating studies, unless their policies were harmonized.

The establishment of the FRC in 1959 was the first attempt to coordinate the policies of these agencies, including those involved in research on the effects of radiation. In a second attempt, the functions and staff of the FRC were transferred to the EPA upon its establishment in 1970. Neither of these efforts at coordination is judged by participants to have been totally successful. The FRC's effectiveness was limited by a policy of obtaining unanimous consent among its members before initiating any action. The EPA stands as one interested agency among many and is burdened with extremely large and distracting additional program responsibilities.

The most recent attempt at coordination stems from the establishment of the RPC to coordinate regulatory activities and the IRRC to coordinate research programs. As noted earlier in this chapter, the agency representation on these two committees overlaps considerably. However, it was believed that different echelons of officials would participate in each of the two groups.

The Interagency Task Force on the Health Effects of Ionizing Radiation, which had recommended the formation of these two interagency groups, had also proposed that both committees be located within the Executive Office of the President in order to elevate their importance within the government and that a minority of their members be representatives of the public in order to enhance the legitimacy of their pronouncements. However, the Executive Order creating them did not give them either Executive Office status or public membership (U.S. President, 1980). It was believed that the Executive Office should not be burdened with additional operational responsibility and that appointment of a minority of public members to the committees would be an inappropriate way to involve the public in the affairs of the agencies.

The RPC is chaired by the Administrator of the EPA, and the IRRC is chaired by the Director of NIH. The 25 agencies represented on one or both of the interagency committees contrast sharply with the initial membership of the FRC, which was limited to only seven agencies.

Congress is considering legislation that would enhance the powers and standing of the committees and require public membership (U.S. Congress, Senate, 1980). For example, the IRRC would be renamed the Federal Conference on Research into the Biological Effects of Radiation, would have its mission broadened to include nonionizing as well as ionizing radiation, would include two public members appointed by the President with the advice and consent of the Senate, would review proposed research studies, and would hold subpoena power. Left unaddressed by the proposed legislation is the coordination of the dozen subcommittees within the Congress that are concerned with formulating and reviewing federal radiation policies and that add to the public's confusion by their frequently contradictory efforts.

The quest for improved coordination of radiation research by the Executive Branch is also the government's response to a more deep-seated dissatisfaction with federal radiation research policies. For many critics the crucial issue is the apparent continued domination of research on the effects of radiation by agencies whose mission it is to promote radiation technologies, especially DOE, which is the linear descendant of the AEC in nuclear affairs. Some officials hope that the integration of radiation research policies in an interagency committee will eliminate the perception of DOE dominance, thereby enhancing the credibility of the government's research efforts. Suggestions to reduce the DOE share of research on radiation effects by transferring funds to the NIH or other agencies and to involve public members in the deliberations of the interagency committees have the same intent.

Although the FREIR Committee recognizes and endorses the virtues of improved coordination among the agencies sponsoring research on the effects of ionizing radiation, it does not believe that reorganization per se will greatly improve the credibility of research results obtained by government agencies. Much of the opposition to DOE and other technology-promoting agencies stems not from the fact that they develop technologies, but from the fact that they develop certain technologies. Objections are voiced by citizens who abhor a dependency on nuclear weapons, the promotion of nuclear power, and the proliferation of certain types of medical devices. The endorsement of research conclusions by an interagency group will not satisfy demands that national policies supporting such radiation technologies be reversed. Still others are concerned about what they perceive to be the refusal of government to provide compensation for individuals putatively exposed to harmful levels of ionizing radiation as the result of agency actions. These individuals are likely to remain skeptical of pronouncements that do not agree with their notions of justice regardless of the level of government from which these pronouncements emanate. The FREIR Committee is doubtful that the proposed structural reorganization would result in a settlement of such intensely divisive political issues.

Indeed, if that reorganization were to produce a recentralization of authority in ionizing radiation research, the committee believes that the government's credibility in this area is likely to be subjected to even further questioning. Public acceptance of the results of scientific research is enhanced when the same results are generated by different groups of investigators that use different approaches and have different sources of support. The existence of the current number of agencies supporting radiation hazard research—some with the mission to promote technologies, some with the mission to regulate technologies, and some with the mission to support general scientific research—may appear disorderly, but it does have the effect of increasing the breadth of reliable information pertaining to the effects of ionizing radiation and providing the opportunity for the development of confirmatory evidence in the field.

Technology-development agencies such as the DOE and the DOD should be encouraged to sponsor research on the risks of exposure to ionizing radiation. These agencies should never be blind to the potential negative consequences of their endeavors. Technology development no less than technology regulation should be informed by results of research on the health effects of ionizing radiation. There are important reasons to separate functions within government, but none to restrict the acquisition of knowledge and sponsorship of research by agencies responsible for the management of the several functions.

Currently, the DOE sponsors slightly less than 50% of the federal research on the biological effects of ionizing radiation. Fifty percent has been discussed as a ceiling for such DOE support. The FREIR Committee sees no special virtue in establishing this ceiling. Forcing agencies to restrict or expand their research activities to meet arbitrary percentage goals is a formula for increasing public suspicions rather than for assuaging them. This seems especially true when a significant portion of the work is likely to remain in the national laboratories because of the uniqueness of their facilities and capabilities, regardless of which agency supplies the funds. A more serious concern is that agencies pursuing their individual responsibilities will ignore or neglect the national need for adequate support of basic research and scientific training to advance knowledge concerning the effects of ionizing radiation. Some argue that this potential lack of support is sufficient justification for the establishment of a national radiation and/or radiobiology institute within the NIH. However, the pervasiveness and variety of interests in radiation makes it unlikely that one or two federal agencies, focused on the medical aspects of radiation, could adequately serve all needs for support. Rather, it would seem that such support is best made a government-wide concern. If medical aspects of radiation require special attention, they might be addressed in programs established within the DHHS to consider the research and training needs of health care technologies, including those pertaining to radiation.

Although the FREIR Committee wishes to see adequate support

for research on the effects of ionizing radiation, it does not believe that this subject should receive a disproportionate share of attention by the government. Consideration of these effects apart from other man-made and natural hazards exaggerates its relative dangers and distorts research and regulatory priorities. The committee recognizes that the public harbors a great fear of the health effects of ionizing radiation, especially its potential for causing cancer and for producing genetic damage. This concern is heightened by the intense and continuing debate over national energy policy, a debate in which some participants have been tempted to resort to the use of unsubstantiated claims about health and safety risks of contending technologies in order to gain advantage for the option they favor. But the government's pandering to these fears hinders the public's ability to appreciate and balance the true risks it faces. For this reason, the committee agrees both with legislative attempts to integrate the coordination of ionizing and nonionizing radiation research (although it proposes that this be done in conjunction with a broader assessment of environmental risks) and the executive decision not to bring this type of research coordination directly into the Executive Office of the President.

The proposals to add public membership to the coordinating bodies stem partly from a belief that scientists alone should not resolve the issues of safety, compensation, and policy direction that beset the studies of the effects of ionizing radiation. Certainly, regulatory decisions and public opinion about what is safe or compensable, and the determination of public policies, are political judgments only partially informed by the current state of scientific knowledge, despite the fact that relatively more is known about the effects of ionizing radiation than most other environmental hazards.

There is a need to force such judgments if public confidence in the effectiveness of government is to be maintained. But the proposals to add public members stem also from a desire to provide forums in which the necessary decisions on radiation safety and compensation are reached. The committee believes that this rationale is inappropriate. The presence of scientists on the committees will not ensure that judgments will be reached on these issues on purely scientific grounds nor will the presence of members of the lay public make them legitimate. Because the important decisions are political, they should be made by politically responsible officials. It is the Courts, the President, and the Congress who must act.

Still, there is a need to coordinate research on the effects of radiation. Gaps in the information base must be identified. Problems common to various agency interests require consideration. Joint resources should be allocated through consultative arrangements. Serious interagency review must be given to major undertakings, especially to the initiation of new epidemiologic studies. The FREIR Committee also recognizes the need to coordinate the regulation of radiation sources. To prevent the further proliferation of federal voices in radiation affairs, it proposes that these functions be combined into a single body, a consolidation of the IRRC and RPC. The FREIR Committee believes that the combined responsibilities are too great for a single participating agency to manage and suggests they be located in an interagency unit to be designated by the President. The agency membership of the new coordination unit should be limited to agencies having important budgetary and legal responsibility in radiation affairs, a number somewhat less than the 25 currently holding memberships on the IRRC and RPC. Agency representation should be at the most senior level, and subcommittees and task forces should be formed to undertake such specific activities as project review or program planning.

Management of Federal Research Program

The committee's review of agency management of research on the biological effects of ionizing radiation focused on the processes by which agencies determine their research strategies, allocate resources to implement these strategies, select specific projects for support, and evaluate the results of research to ensure maximal return on their research investment. Several deficiencies were observed. Some of them are correctable through improvements in agency practices. Others require the attention of an interagency coordinating mechanism. Still other deficiencies are inherent in the operations of a complex government; at best, only some of their consequences can be mitigated. The committee was favorably impressed with the overall quality of federally supported research in the field of ionizing radiation. This good quality is perhaps the result of the priority attached to nuclear-related activities after the Second World War. The field has been well supported until recently, and there has been much progress. In comparison to most risks encountered by individuals and society, an extraordinary amount is known about the effects of ionizing radiation on human health.

There are, however, limitations to the progress that can reasonably be expected in the near future, particularly in gaining a full understanding of the health effects of exposures to low levels of radiation.

No methodology appears likely to provide clear, unique answers in this important area of investigation. More funding alone will not provide the answers needed, nor will changes in management practices. Simply stated, there are research goals—the determination of the health effects of exposures to low level radiation may be among them—that exceed the currently realized and envisioned capabilities of science.

Apparently, the public has little appreciation of these limitations. On the contrary, the public places great pressure on political leaders to assure the absolute safety of radiation technologies, reacting emotionally to every reiteration of their potential hazards. In turn, political leaders pressure agencies to produce immediate and definitive statements of the risks involved. Too often the response results in the initiation of studies that are unlikely to yield meaningful results. In the aftermath of the Three Mile Island Nuclear Power Plant accident, for example, there were the inevitable calls for epidemiologic studies of the affected populations. These studies were initiated despite the fact that the levels of exposure were such that the demonstration of biological effects relating to the exposure is virtually impossible.

If the public is to be better informed of the actual risks it faces and of the processes by which scientifically valid conclusions are reached, it is incumbent upon agencies to resist sponsorship of projects that are inherently flawed. The political choice, nevertheless, might be to proceed with a study; there are legal and other reasons to gather data that are of limited scientific utility. If that is the case, the government should strive to distinguish such an effort from a scientifically valid investigation. For any element of government to pretend otherwise is to harm the interests of both the public and science.

Much of the work on the effects of ionizing radiation, especially studies of laboratory animals and human populations, requires many years of effort to produce meaningful results, necessitating the commitment of long-term support. The vagaries of the budgeting process, however, are such that agencies are reluctant to allocate their limited resources to long-term commitments: priorities may shift, new projects may be assigned without additional resources, and government-wide cutbacks or congressional directives may eliminate agency flexibility. The committee believes it is imperative that the proposed interagency coordinating committee ensure that sufficient resources be set aside either by individual agencies or several agencies through sharing agreements to provide the support required for continuing projects in this field.

A complication for long-term studies is the difficulty of maintaining quality. The work is often also tedious and mundane. For example, animal colonies must be maintained for long periods and statistics on individuals for epidemiologic studies must be gleaned from scattered public health and employment records. Most scientists, oriented as they are toward the rewards of rapid publication of research results, find little advantage in such activities. Appropriately, much of this work is conducted in the national laboratories managed by the DOE. However, like other federal agencies, the DOE is forced to shift resources rapidly from one topic to another in response to the changing national political agenda. The facilities and staff necessary for the long-term support of research in ionizing radiation hazards are seriously threatened by these priority shifts. Again, the committee believes it is imperative that the proposed interagency coordination committee address itself to this problem and seek to guarantee that adequate resources be provided to the national laboratories and universities for vital long-term studies on the effects of ionizing radiation.

The national laboratories and universities, of course, are not immune from the effects on morale resulting from the performance of work that, although necessary, is often repetitious and unchallenging. Staff assigned to these tasks can easily feel scientifically isolated. Recruitment and retention problems are often a result. The DOE is aware of these problems, but it should be encouraged to do more than it has to ensure that the quality of the laboratory staffs is maintained. Rotation of assignments, sabbaticals, and research opportunities are techniques that could prove to be useful. Especially effective is the involvement of an investigator in both long-term and related short-term studies simultaneously. The short-term studies generate new ideas, maintain competence, and increase the opportunity to publish scientific papers.

Data gathered through long-term studies are costly to acquire and difficult to replicate. The data should be carefully collated and stored so that they may be reanalyzed to reexamine findings or to apply new techniques as methodologies are improved. Unfortunately, too little care is taken by agencies to acquire and store data from studies they have sponsored. The proposed interagency committee ought to establish standards for the retention of research data and procedures for making the data available to groups both within and outside of government. The rights of investigators must be protected, but so too must be those of society. The cost and time penalties imposed by the loss or unavailability of data greatly hinder progress in improving our understanding of the health effects of ionizing radiation. Moreover, the need to repeat studies may involve placing humans or animals at risk unnecessarily.

No agency currently has the mission of ensuring that an adequate supply of scientific and technical manpower with appropriate skills is available to study the effects of ionizing radiation. The AEC once had this responsibility, but with reorganization it has been lost. The large number of skilled personnel who had been nurtured by the AEC are now retiring from research. Without replacements, the nation may not have the necessary expertise to respond to the inevitable crises and to maintain the required continuing studies. DOE, NIH, and NSF should prepare a joint program to assure the adequacy of the manpower pool in this field, including, if necessary, the provision of training support.

The committee found that scientists were often poorly informed of or confused by the procedures used by the agencies to select projects. For example, there was a surprising difference between the descriptions of agency policies provided by the agencies and the perceptions of these policies by scientists who were current or potential recipients of agency support. To be sure, inevitable disagreements in these perceptions may result from the rejection of specific projects by the agencies on the basis of quality or priority. But it is also clear that agencies have failed to inform scientists adequately of current priorities and standards. Since the scientists will conduct the research and since they must make their own plans, this is a serious deficiency.

The agencies also differ markedly in their project selection practices. At the extreme, some agencies rely almost totally on program managers to determine the projects to be included in their research portfolios, while others rely almost exclusively on panels of outside experts for project selection. Both systems have strengths. Program managers are closely tied to agency missions and can use a variety of mechanisms, e.g., project grants, contracts, and center grants, to obtain their objectives. Panels permit the agencies to rely on recognized experts in defining the content of their programs. But there are problems with each of these approaches.

The committee is concerned that agencies relying on their program managers fail to provide sufficient external review of the research projects they select for support. The failure to involve outside experts in the selection process undermines the credibility of the agency and gives rise to the appearance that the agencies are selecting research programs whose results are likely to be compatible with official policies. In a field as socially sensitive as research on the effects of ionizing radiation, there is a need for agencies to seek not only the substance but also the appearance of total objectivity. Peer review is one of several procedures by which the scientific community imposes the norm of objectivity upon the selection of research projects. To be sure, controversy is inevitable in the field of ionizing radiation as in some other fields, given both the limitations of the current state of knowledge and the political consequences of research results. Nevertheless, a sincere attempt to ensure objectivity in project selection would buttress agency arguments that they are concerned with the discovery of truth as well as with the pursuit of their specific technological missions.

The committee is aware that the peer-review system is not flawless. Scientists are worried that there is a tendency for the review to become overly conservative, especially as the gap grows between requests for support and available funds. Thus, innovation in science is at risk. Moreover, there is a potential conflict between the desire for broad representation on the panels and the need for maintaining the highest quality in the reviews. The influence of reviews on the distribution of public funds makes it inevitable that the process for selecting reviewers will be subject to public scrutiny and the application of social criteria. Democratizing the selection of reviewers, however, does not guarantee that either science or the public good will be advanced.

In fiscal year 1981, the federal government expects to spend $111 million on research on the health effects of ionizing radiation (see Table 9-1). Most of this work will be sponsored by DOE and NIH, but 13 other agencies will contribute as well. Sixty percent of the research will be conducted by the national laboratories and other government research facilities. The prime emphasis is to be placed on animal models and epidemiologic studies.

The committee believes that the current level of support is appropriate. However, it also believes that more emphasis should be placed on basic science investigations, especially those in cell biology. Such work holds the greatest promise for deepening our understanding of the effects of exposure to low levels of ionizing radiation. In the field of epidemiology, the committee proposes that support be focused on the improvement of investigative techniques, e.g., the use of markers, rather than the initiation of additional large-scale studies. It also believes that more attention should be placed on exploring the extrapolation of data from studies of nonhuman systems to humans. Until there is significant progress in the advancement of investigative techniques for measurement, it is doubtful that there will be important advances in understanding the health effects of exposures to low levels of ionizing radiation.

Public Information

A responsibility of government in sponsoring research on the health effects of ionizing radiation is to inform the public of the risks and benefits of radiation exposures. It is the government's obligation to distribute accurate and timely information on this subject regardless of the impact on current official policies.

Surprisingly little is known, however, about the public's attitudes toward radiation and the role that government information programs play in the formation of those attitudes. This is true even if the inquiry is broadened to include public attitudes toward nuclear power generation, a highly visible and controversial application of radiation technology. The public is sharply divided on the issue of nuclear power development, its conflicting opinions are held intensely, and the opinions of the antinuclear advocates reflect a variety of fears concerning the health effects of radiation. It is not known with any degree of certainty how these opinions developed or the degree to which they can be modified by additional knowledge. Opinion surveys have failed to probe the dynamics of these public attitudes in any depth.

There is some evidence that technical and scientific information, even when it has been made available, has little effect on these strongly held opinions. The public, it would appear, is bewildered by the language of radiation science. For example, the distinction between a rem and a millirem, or between a curie and a picocurie, will not be readily apparent to most people.

There may also be a major problem of credibility. Over the years, government spokesmen have not always been forthright in their reporting to the public on radiation matters. The threat of massive claims for compensation against certain federal agencies further tends to undermine the authority of government pronouncements on nuclear matters in general. In the post-Watergate atmosphere, many Americans appear to assume that public agencies and public officials are not above tampering with scientific evidence. Whether or not it is understood or even believed, however, the government has no choice but to continue its efforts to disseminate information.

Raw data can be misleading. In the committee's judgment, government agencies must try harder to present research findings in their appropriate context, to explain the scientific processes by which the information was obtained, and to clarify the significance of the implied risks. Officials may have only themselves to blame when they rush into print with alarming projections of dangers that upon closer examination turn out to be more apparent than real. The provision of context and perspective may be the most important contribution government can make. A truly informed public must be able to discriminate among various interpretations of the same set of facts and to appreciate the uses of scientific knowledge and the limits of certainty as applied to radiation.

The committee, of course, is aware of the skepticism with which government pronouncements in this and other areas of public concern tend to be received. In the current climate, the disclosure of newly discovered information that appears to contradict previously announced findings can be perceived as an attempt at official deception. Products once certified as safe may, in the light of fresh evidence, have to be labeled unsafe. Reversals of this kind are consistent with the evolution of new knowledge and with the responsibilities of public agencies. Such reevaluations should not be confused with deliberate misrepresentation.

For these reasons, the committee cannot support the establishment of a single national agency to disseminate information on radiation hazards. The idea that such an agency might be established can be inferred from the report of the Interagency Task Force on the Health Effects of Ionizing Radiation (Department of Health, Education, and Welfare, 1979). The committee believes that the public interest will be better served by a diversity of agencies engaged in radiation research, each one encouraged to publish details on its individual approaches and findings. The appearance of contradiction, when that occurs, is preferable to a facade of enforced unanimity. The standing of each agency within the scientific community can only be enhanced by the seriousness with which it fulfills its obligation to inform the public.

The Use of Research Results

Contrary to what might be assumed, there are essentially no direct pathways by which research results find their way to the government officials responsible for setting radiation protection standards and regulatory policies. Instead, results of research are communicated from the scientific community to the federal agencies needing it through a number of channels outside of the government that have become established over many years. One of these is the open scientific literature in which research investigators publish their findings regularly. Those who are responsible for radiation standards and regulatory policies are expected to keep abreast of this literature. Research results that are buried in technical reports or otherwise confined to a limited distribution are clearly of less value than those appearing in more easily accessible publications.

There are two other pathways by which research results on the health effects of ionizing radiation may be transmitted to those needing them: the reports of the National Council on Radiation Protection and Measurement (NCRP) and the reports of the National Academy of Sciences/National Research Council (NAS/NRC). These nongovernmental organizations have for many years systematically drawn together able research scientists within this country to review the scientific literature from time to time and to write reports on an ever-expanding series of topics pertinent to protection from ionizing radiation. The NCRP reports provide detailed information concerning safe operating practices as well as recommendations pertaining to radiation protection standards. NAS/NRC reports, through comprehensive analyses of the scientific literature, have provided many useful summaries and interpretations of the health risks of ionizing radiation. Together, the NCRP and NAS/NRC provide another valuable feedback loop, transmitting research data to agencies concerned with the control of ionizing radiation.

REFERENCES

Department of Health, Education, and Welfare. 1979. Report of the Interagency Task Force on the Health Effects of Ionizing Radiation. Office of the Secretary, U.S. Department of Health, Education, and Welfare, Washington, D.C. 113 pp.

National Academy of Sciences. 1972. The Effects on Populations of Exposure to Low Levels of Ionizing Radiation. Report of the Advisory Committee on the Biological Effects of Ionizing Radiations. National Academy of Sciences, Washington, D.C. 217 pp.

United Nations. 1969. Report of the United Nations Scientific Committee on the Effects of Atomic Radiation. Official Records of the General Assembly: Twenty-Fourth Session, Suppl. No. 13 (A/7613). United Nations, New York.

U.S. Congress. 1979. S. 685, a Bill to Establish a Program for Federal Storage of Spent Fuel from Civilian Nuclear Powerplants, to Set Forth a Federal Policy and Initiate a Program for the Long-Term Storage of Nuclear Waste from Activities. 96th Congress, 1st Session.

U.S. Congress, Senate. 1980. Federal Radiation Protection Management Act of 1980. Senate Report 96-925. Committee on Governmental Affairs. 96th Congress, 2nd Session.

U.S. President. 1970. Environmental Protection Agency. The President's Message to

Congress Transmitting Reorganization Plan 3 of 1970. July 9, 1970. Weekly Compilation of Presidential Documents 6(28):917-919.

U.S. President. 1980. Radiation Protection Program. Announcement of New Initiations. February 21, 1980. Weekly Compilation of Presidential Documents 15(8):364.

U.S. Public Law 83-703. 1954. The Atomic Energy Act of 1954. 42 U.S.C. 2011 *et seq*.

U.S. Public Law 86-70. 1959. Federal Water Pollution Control Act. 33 U.S.C. 1157 *et seq*.

U.S. Public Law 86-373. 1959. To amend the Atomic Energy Act of 1954, as amended, by adding 274 with respect to cooperation with states. 42 U.S.C. 2021.

U.S. Public Law 88-206. 1963. Clean Air Act. 42 U.S.C. 1857 *et seq*.

U.S. Public Law 93-438. 1974. Energy Reorganization Act of 1974. 5 U.S.C. 5313; 42 U.S.C. 5801 *et seq*.

U.S. Public Law 93-523. 1974. Safe Drinking Water Act. 88 Stat. 1660. U.S. Public Law 94-580. 1976. Resource Conservation and Recovery Act of 1976. 90 Stat. 2795.

U.S. Public Law 95-91. 1977. Department of Energy Organization Act. 91 Stat. 565 *et seq*.

U.S. Public Law 585. 1946. The Atomic Energy Act of 1946. 42 U.S.C. 1801-1819.

11
Future Research and Its Management

During the past half century, federally supported research has provided a vast body of knowledge on the biological effects of ionizing radiation. Current research, reviewed in Chapter 9, constitutes but a small increment of a much larger investigative effort. As in all scientific endeavors, much remains to be learned. However, this should not distract us from the fact that there is a substantial body of scientific information on radiation effects.

At this point, it is appropriate to ask the following questions:

- What directions should federally supported research take in coming years to assist those who are responsible for the control of ionizing radiation?
- What research is needed to improve estimates of radiation risk?
- What research is needed to enhance benefit and diminish risk in technologies using ionizing radiation?

An easy answer to all of these questions is to suggest that future research merely continue work currently in progress, perhaps expanded somewhat in scope and financial support. But such an answer is quite inappropriate because it fails to recognize that public attitudes and perceptions with respect to ionizing radiation have been changing and that these changes may require future research to follow important, additional pathways. Clearly, these questions require careful examination.

Future Research

To provide an answer to the first question, let us examine briefly the processes by which decisions are made by federal authorities when responding to problems in the field of ionizing radiation. Currently, responsibility for the formulation and promulgation of radiation protection standards resides within the Environmental Protection Agency (EPA). Decisions pertaining to regulatory policies for the nuclear industry are the responsibility of the Nuclear Regulatory Commission (NRC), and, for medical uses of ionizing radiation, they fall within the purview of the Bureau of Radiological Health (BRH). In all three agencies, actions are taken on matters brought before them only after careful consideration has been given to a broad range of factors. Scientific information is, of course, one of these, but social, economic, and political factors also play a decisive role, sometimes predominantly so. Moreover, although much remains to be learned about the health effects of ionizing radiation, scientific information is sufficient to permit federal authorities to formulate radiation protection standards and to delineate comprehensive regulatory policies without serious limitation. The EPA, NRC, and BRH cannot await new scientific information before taking decisive action on matters within their several responsibilities.

From the foregoing, it may be inferred that current research on the biological effects of ionizing radiation is no longer of vital interest to the several regulatory agencies. Although it may be true that officials of these agencies are only interested bystanders observing scientists wrestling with such matters as the dose-effect relationships of ionizing radiation, a number of facets of current research have a major bearing on many of their actions, such as the decision to fund research to devise models for predicting radionuclide distributions in the environment and in humans. Such models are essential for determining radiation doses received through radionuclides entering the body. Hence, they play an important role in the development of certain radiation standards and policies. Also, there is no question about the usefulness of research to improve dosimetric instrumentation. Radiation standards and policies are dependent upon the availability of a broad range of appropriate options for use by public officials and others involved in the applications of ionizing radiation.

The issues are not as clear with respect to future epidemiologic studies and to research on animals, lower life forms, plants, cells in tissue culture, and biological substrates undertaken to gain better understanding of radiobiological mechanisms and dose-effects relationships. Since results of such research will tend to emerge slowly, they are not likely to have major day-to-day impacts on the affairs of the EPA, NRC, and BRH. Justification for continuation of these investigations must therefore be found elsewhere.

It was pointed out in Chapter 4 that estimates of the health risks of ionizing radiation are based primarily on observations made in several large populations exposed to relatively large radiation doses. There have not been sufficiently large populations of humans exposed to accurately recorded low doses to yield statistically reliable data on the biological effects of ionizing radiation. As a consequence, estimates of human risk at low doses have frequently been derived from linear extrapolations of high-dose data to zero dose levels. For high linear energy transfer (LET) radiations, such a process probably estimates risk within reasonably accurate limits; for low-LET radiations, however, such extrapolation may overestimate risk (National Academy of Sciences, 1980). Because the vast majority of individuals living within the United States are likely to be exposed to relatively low doses of ionizing radiation that are low-LET in character, it is clear that additional research is needed to improve risk estimates of low-LET radiation at low doses (25 rem and less).

Because experiments cannot be conducted in humans, such research must be focused on controlled laboratory studies in which effects are observed in animals, lower life forms, plants, cells in tissue culture, and biological substrates. Since these observations are not made in humans, their applicability in determinations of human risk is subject to question. Such uncertainties may be expected to disappear only when future research leads to an understanding of the principles of biological damage such as carcinogenesis and repair in all animal species, including humans. Hence, future research to improve estimates of risk to humans exposed to radiation must take its place as a part of the broad investigative effort now in progress in the United States to discover the basic principles of cancer formation and other biological aberrations, such as genetic damage. Moreover, proposals for such research support should be judged not on their significance within the radiation sciences alone but in competition with proposals seeking fundamental knowledge in the biological sciences generally.

Future research to improve estimates of radiation risk should include additional epidemiologic research. In the past, much has been learned from the epidemiologic studies of such population groups as the survivors of the Japanese bombings, the uranium mine workers, and several groups of patients in whom x-rays have been used diagnostically and therapeutically. Because such studies by their very nature extend over long periods, many of them are still in progress. These should be continued with periodic peer review until they have reached their logical conclusions.

New epidemiologic studies, however, should be undertaken only with great care. From time to time, there will undoubtedly be populations in which exposure to ionizing radiation has occurred and which for political reasons may seem attractive for intensive study. Seldom, however, will these populations be sufficiently large, nor will their radiation doses be adequately documented to yield statistically significant data on dose-effect relationships and radiation risk. The population residing in the vicinity of the Three Mile Island nuclear power facility is such an example. This population may well have been well suited for quantitative studies of psychoneurological effects following a nuclear accident. However, this population was much too small for an epidemiologic study of dose-effect relationships. Furthermore, radiation doses for much of that population were much too small to yield statistically significant results. Regardless of the social and political pressures to do otherwise, such epidemiologic studies cannot be justified on scientific grounds.

The committee has found that there has been relatively little support for research to enhance benefit and diminish risk in technologies using ionizing radiation. The BRH has a modest program dealing with radiation dose reduction in diagnostic medical procedures. Beyond that, only limited funding had been allocated for research to develop medical technologies that increase the quantity of diagnostic information while maintaining or reducing radiation dose. The committee believes that there should be greater support for studies in this area.

Among the social, economic, and political factors that are important in the setting of radiation standards and the development of regulatory programs is the public's attitude toward radiation risk. Currently, there is a substantial dichotomy in this attitude. On the one hand, the public appears to accept the radiation risks associated with exposure to natural background and medical sources (combined lifetime doses averaging approximately 15 rem). On the other hand, it has not done so with respect to the risks associated with facilities of the nuclear industry, especially those for generating nuclear power. After more than two decades of experience with such facilities, major segments of the public remain skeptical of their safety.

Data compiled by the Committee on the Biological Effects of Ionizing Radiations (the "BEIR Committee") indicate that the annual whole-body dose of ionizing radiation received by the public from nuclear power facilities is less than 1 mrem, i.e., a dose less than 1% of that received from medical and natural radiation sources (National Academy of Sciences, 1980). Even in the vicinity of a malfunctioning power reactor such as that at Three Mile Island, radiation doses of the nearby population have not exceeded 1 mrem (President's Commission on the Accident at Three Mile Island, 1979). Clearly, the public's attitudes and perceptions regarding the risks associated with the nuclear industry stand in sharp contrast to those pertaining to medical applications.

Because an understanding of the nature of risk perception is important to the legislative and executive branches of the federal government responsible for decisions relating to energy and science policy, it is important that the question be addressed.

Management

During the past 10 yr, the administration of federally supported research on the biological effects of ionizing radiation has undergone considerable fragmentation. Many agencies are now involved in these programs, whereas during the 1950's and early 1960's the great majority of such research was supported by the then-existing Atomic Energy Commission.

This fragmentation does not appear to have been detrimental to the quality or conduct of the research. On the contrary, it may have been beneficial by providing multiple focal points for different interests and emphases and greater opportunities for funding a wide variety of research.

The fragmentation also does not appear to have been detrimental to the management of radiation research. Coordination of the various agency programs may present only minor problems at the moment, but there is reason to believe that an Interagency Committee on Radiation Research, created by presidential executive order, will be successful in maintaining an orderly development of the several radiation-related research programs sponsored by the federal government.

What seems to be needed now, after the major changes that have occurred during the past decade or so, is an opportunity for the several federal agencies that have been given responsibility for the administration of research on biological effects of ionizing radiation to carry out their functions without organizational disruption and distraction. If this is done, the radiation research programs of the future will probably be as productive and distinguished as those of the past.

References

National Academy of Sciences. 1980. The Effects on Populations of Exposure to Low Levels of Ionizing Radiation. Report of the Committee on the Biological Effects of Ionizing Radiations. National Academy of Sciences, Washington, D.C. 638 pp.

President's Commission on the Accident at Three Mile Island. 1979. Staff Reports to the President's Commission on the Accident at Three Mile Island. U.S. Government Printing Office, Washington, D.C.